Hawaiian Volcanoes

Hawaiian Volcanoes

Clarence Edward Dutton

Foreword and Appendixes by

William R. Halliday

A Latitude 20 Book

University of Hawai'i Press
HONOLULU

Printed in the United States of America

05 06 07 08 09 10 6 5 4 3 2 1

Originally published by the Department of the Interior in
the 4th Annual Report of the U.S. Geological Survey, 1883

Library of Congress Cataloging-in-Publication Data

Dutton, Clarence E. (Clarence Edward), 1841–1912.
Hawaiian volcanoes / Clarence Edward Dutton ; foreword and appendixes
by William R. Halliday.
p. cm.
"Originally published by the Department of the Interior in the 4th Annual
report of the U.S. Geological Survey, 1883"—T.p. verso.
Includes bibliographical references.
ISBN-13: 978-0-8248-2960-5 (acid-free paper)
ISBN-10: 0-8248-2960-3 (acid-free paper)
1. Volcanoes—Hawaii. 2. Dutton, Clarence E. (Clarence Edward),
1841–1912—Travel—Hawaii. 3. Hawaii—Description and travel.
I. Halliday, William R., 1926– II. Title.

QE524.2.U6D88 2005
551.21'09969'1—dc22
2005043998

University of Hawai'i Press books are printed on
acid-free paper and meet the guidelines for permanence
and durability of the Council on Library Resources.

New material designed by Lucille C. Aono
Printed by Edwards Brothers, Inc.

Contents

A note on pagination: This report was first published as part of a larger volume and the original pagination has been retained. For this reason the text begins on page 80. Also, Chapter XI (pages 183–198) has been omitted from this reprinting.

Foreword

Clarence Dutton's Hawai'i, Then and Now

"Hawaiian Volcanoes" is the second of Clarence Dutton's memorable 19th-century geological travelogs reprinted by a 21st-century university press. His earlier monograph on the Grand Canyon area was acclaimed immediately. It still is commonly credited with popularizing the sublimity of that awesome landscape.[1] When he came to Hawai'i he was confronted by another world wonder: the orange-red luminosity of Kīlauea volcano's cauldrons of churning lava. Again his unique command of the English language rose to the challenge. In 2002, an authoritative web site of the prestigious American Geological Institute described it as "beautifully written"—a remarkable phrase for a formal report of the U.S. Geological Survey in any century. But "Hawaiian Volcanoes" was, and is, much more than a mere geological report. It takes the form of an entrancing "roadside geology" of Hawai'i's Big Island and much of Maui, a century ahead of its time.[2] And whether incidentally or intentionally, it also was a subtle socioeconomic analysis of human conditions at an especially turbulent moment for the shaky Kingdom of Hawai'i. The reader is left with a nagging question: Was Dutton's assignment to Hawai'i purely geological?[3]

Dutton's Literary Genius

Dutton possessed a painter's eye for color, light and shade, and admirable form, and an architect's eye for line and proportion. His literary genius was to express these with clarity of word, phrase, and paragraph. This gave his writings a peculiarly visual, highly pictorial quality that allows the landscapes to unfold for the reader as they do for the traveler. Through-

---

[1] In the foreword to the 2001 University of Arizona Press reprint of *Tertiary History of the Grand Canyon District*, Stephen Pyne wrote that no one before Dutton saw what he did, and, since then, no one has been able to see anything else. He was preceded by Wallace Stegner (1953) who wrote that it is with Dutton's eyes that the ordinary visitor views the vast canyon.

[2] A preliminary version of *Tertiary History of the Grand Canyon District* appeared as "The Physical Geology of the Grand Canyon District" in the 1882 *2nd Annual Report of the U.S. Geological Survey*, pages 47–166.

[3] Dutton's puzzling assignment to Hawai'i is discussed in detail in Appendix A.

out his works he wrote primarily as an enthusiastic geologist, not as an educator or entertainer. He often subordinated literary skills to acute observation and inspired interpretation. Consistently, however, he embedded technical facts in a highly readable form that sweeps the reader along like the chapters in a serial romance. Thus his works are deceptively simple and his observations meticulously accurate despite their eminent readability. Accuracy, charm, and completeness are manifest so uniformly that their rare absence provokes careful reflection. Painstakingly he differentiated observations from interpretations, and both from speculations impelled by the primitive state of geology in 1882. To the reader familiar with the scenes he described, nothing seems exaggerated. He focused his frequent enthusiasms proportionately, from the merely interesting to the most remarkable or sublime. His occasional lapses into the gloomy "pathetic fallacy" beloved of writers of his Victorian period always were brief and serve as effective contrasts to the glorious reality on which he focused.[4] Seemingly all that is missing is Dutton's persona. "Hawaiian Volcanoes" contains a little more about his personal feelings (other than for geology and scenery) than do his major writings about the western United States. But Dutton despised publicity seekers so thoroughly that his writings ruthlessly suppressed the humanness of the extraordinary man behind the prose. As a result, no adequate biography exists for Dutton. This account is the first to consider why he went to Hawai'i (other than as a vacation, which it certainly was not) or even the influence of the world leaders with whom he associated at the Cosmos Club, 6 months a year for 15 years. Late in life, his son recalled that Dutton could play seven games of chess simultaneously while blindfolded. Clearly he was a genius in many fields, and literature was not the least of them.

Dutton's Beginnings of the Great Tour: Chapters I and II.

Near the southern tip of Hawai'i Island, Dutton began his account at little Wai'ōhinu, beloved of Mark Twain. There he "fitted out a pack train with six packs in Rocky Mountain style." With a packer, cook, and the first of his guides, he rode off into Paradise. Immediately the reader is caught up in his eager enthusiasm and lyrical prose:

> In all those portions of the islands where [volcanic] quiet has long prevailed, the scenery has habitudes of extreme boldness and animation. It is always picturesque, very often beautiful, and sometimes grand. Cliffs, crags and canyons are carved in the mountains with as much sharpness and spirited

[4] The "pathetic fallacy" consisted of description of unpeopled scenery with brooding language now largely reserved for haunted houses and Gothic novels.

> detail as in the Plateau country. In west Maui and Kauai may be found walled valleys and amphitheaters that almost rival Yosemite. . . .

Terming Mauna Loa "the king of modern volcanoes," he placed it in sprightly perspective; "Mounts Shasta, Hood and Ranier [*sic*], if they were melted down and run together, would fall much below [its volume]." From his ebullient prose, Mauna Loa, Mauna Kea, and Kīlauea immediately take on the individual personalities we attribute to them today. And perhaps above all, "Relatively to human comfort the climate is perfection."

The first chapter is simultaneously an introduction to Hawai'i and a summary of its geology—partly his personal observations and partly those of many who had preceded him. In perusing it, the reader should keep in mind the primitive state of geology in 1882 as well as the strengths and weaknesses of Dutton's unique literary style. The science of geology has moved on in channels scarcely imaginable in 1882. In the normal way of science, some of his conclusions inevitably have been superseded by the weight of later research and observations. Today it is the contagious enthusiasm of Dutton's writing that we admire. Stripping away all the science—Dutton's own observations, the conclusions he drew from them, and the subsequent theorizations—this volume still would remain a perpetual hymn of praise to the islands.

Dutton's fluent style emphasized clarity and cohesiveness above documentation. Halting a flow of words every few sentences to give credit to each source is deplorable writing. Yet it inevitably exposes the writer to charges of ingratitude—or worse. James Dana's 1890 *Characteristics of Volcanoes* dealt almost exclusively with Kīlauea and Mauna Loa (or "Mt. Loa," as he called it).[5] Among geologists it is better remembered than "Hawaiian Volcanoes" because of its careful crediting of even minute matters—as well as the superior reputation of its academic author and his correction of inevitable errors by Dutton. But seemingly half of its considerable bulk is a bone-dry recitation of who did what and when, as tiresome to read as "Hawaiian Volcanoes" is fun. Dutton chose the other path, and we are grateful.

Yet it would be a mistake to wholly overlook Dutton's scientific content. He differentiated rough *'a'ā* lava from smooth *pāhoehoe* with such clarity that in 1993, British volcanologist David Chester credited this account with introducing these two Hawaiian words into the accepted language of modern geology. Plate VI is a notable depiction of freshly glistening *pāhoehoe*, worth many paragraphs of description. Further, in recording that a

[5]The full citation is James D. Dana, 1891. *Characteristics of Volcanoes with Contributions of Facts and Principles from the Hawaiian Islands.* New York, Dodd, Mead & Co. Its Hawai'i sections are derived almost exclusively from Kīlauea and Mauna Loa volcanoes. In 1882 Dana was professor of geology at Yale University and author of a very influential geology textbook as well as *Dana's System of Mineralogy*.

single tongue of basaltic lava can change from *pāhoehoe* to *'a'ā* and back again, he was more than a century ahead of his time. He described the drainage of Mauna Loa in elegant language, revealing its similarity to that of cavernous limestone areas—karstic terrains—he had visited in the eastern United States. (Twenty-five years later the German geologist Walther von Knebel observed similar volcanic terrain in Iceland and dubbed it "pseudokarst.") Dutton somehow misstated the location of Kapāpala Cave but explicated the great earthquakes and mudflows of that region in terms presaging his later impact on American seismology. And he even recorded that the "Great Crack" had been steaming ever since 1838. (It has cooled down since then.)

The thoughtful reader soon will notice seeming inconsistencies in Dutton's appraisals of the *kama'āina* (Native Hawaiians). In Chapter I he announced that "In 25 years after the landing of the missionaries [1820] the whole people had, in a great measure, become Americanized." Facts carefully related in later chapters clearly belied this seemingly peculiar conclusion. Obviously, literacy in the Hawaiian language hardly was evidence of Americanization. But a subtle reason existed for this odd sentence: Back home in Washington, D.C., at least one political faction had urged annexation of the islands to "Americanize" the Hawaiian people. Without ever mentioning the word "annexation," Dutton neatly drew the reader to a carefully planned conclusion: it was not appropriate. "Large amounts of American capital have been invested in the [sugar] plantations and in the accessory commerce, with great advantage to both countries." Throughout the book, he carefully avoided all mention of members of the unstable government except nonpolitical "Professor W. D. Alexander, the Surveyor-General of the Hawaiian Kingdom." Today, this volume is virtually the only convenient source of Alexander's wonderful early maps of Hawai'i, and a few rare British Admiralty charts are included for good measure.

Dutton's sprightly roadside geology of Hawai'i began in earnest in Chapter II. Separately, he ruefully explained his choice of Wai'ōhinu over Hilo as a starting point: "The field geologist quickly gets accustomed to every inconvenience and discomfort of travel except one, and that is mud, and the more he has to do with mud, the more he hates it."[6] Rain reigns in Hilo, often by the proverbial bucketful. Dutton had to visit Hilo eventually in order to ascend Mauna Kea. But the southern route from Wai'ōhinu and Kīlauea summit was dry almost all the way to Hilo. Moreover, it had other advantages. Forty years earlier, the Wilkes party almost came to disaster on Mauna Loa from employing a braggart guide at the Hilo end

---

[6]Early in 1883, Dutton wrote a personal letter to James Dana, who published it in *American Journal of Science* (1883, 3rd Series, 25:219–226) with the title "Recent Explorations of the Volcanic Phenomena of the Hawaiian Islands." Although common early in the 19th century, this practice was decidedly unusual by 1883.

of the trek.[7] Dutton had no such problem. Passing through the Kaʻū District, he had ample opportunity to talk with ranchers and *kamaʻāina* who lived and hunted on its slopes and knew the ancient trails. As will be seen, their knowledgeable guidance enormously simplified Dutton's later work on the vast mountain.

Almost at once in Chapter II, Dutton compares the terraces on Mauna Loa to those of his beloved plateau country of the West. Some of these terraces are of special interest in themselves; they are remnants of Nīnole, a remarkable volcanic structure considered by some geologists to be an older volcano overwhelmed and largely buried by Mauna Loa lavas. (Others consider it to be merely an early feature of Mauna Loa itself.) But Dutton primarily was using this comparison as a literary figure. So rarely did he make such comparisons between Hawaiian volcanoes and features of the plateau country that they underscore the dissimilarity of the two terrains.

The lengthy footnote on the Hawaiian language (as then spoken on the Big Island) is a valuable reminder of how rapidly that language was changing in the 19th century—as it continues to change today. Between Nāʻālehu and Punaluʻu, Dutton's course was along what now is Highway 11. Somewhere around Pāhala, it swung inward from the present highway, via the present Kapāpala Ranch headquarters, to parallel the bare lava fields of the vast southwest rift zone of Kīlauea. Eventually his path must have swung eastward to avoid the enormous Keʻāmuku *ʻaʻā* lava flow descending from high on Mauna Loa, then followed one of several Kaʻū Desert trails that then converged at the summit of Kīlauea. Following his account, the reader is jolted by the sudden virtual view into the

[7] In the tradition of British captains Cook and Vancouver (and with perhaps a touch of Captain Bligh in addition), Charles Wilkes commanded the U.S. Exploring Expedition of 1838–1842 in what then was called the South Seas. It also included the coast of northwestern America and what became Wilkes Land in Antarctica. Broadly talented and educated, Wilkes was a member of the American Philosophical Society and other learned groups. James Dana (see note 5) was one of the scientists in his entourage, and he prepared a volume on his geological observations. The reports of the expedition were meticulous and voluminous, appearing in several volumes—the number of volumes depending on the publisher, as the first edition was pirated repeatedly in those precopyright days. In the 1845 Lea and Blanchard (Philadelphia) 5-volume version, Wilkes wrote half of Volume 4 on his observations in Hawaiʻi as part of his narrative of the expedition. Dana added other observations in Volume 5: Geology. Among his Hawaiian observations, Wilkes perhaps is best remembered for his gravimetric studies atop Mauna Loa. He also was the hot-tempered naval officer who almost embroiled Great Britain in the American Civil War, on the side of the Confederate States of America. Contrary to international law, he stopped a neutral British passenger ship and seized two Confederate diplomats; President Lincoln had to give them back. Among his other writings was a Dutton-like account of "Western America, including California and Oregon." In retirement he was promoted to rear admiral.

caldera, closely akin to the awe and wonder upon first reaching the rim of the Grand Canyon itself. Dutton undoubtedly planned it that way. Few others could have created such a visual impact from written words.

Dutton's View of Kīlauea Volcano and Its People: Chapters III, IV, and VI

Chapter III is perhaps Dutton's most important contribution to Hawaiian literature. Its twin themes are awesome beauty and the constant change of volcanoes in action. The active Kīlauea caldera he described differed greatly from the placid, seemingly monotonous gray summit expanse of today. His selection of earlier accounts and important maps dramatically documented the first half-century of its written history. The foldout "View of Kilauea Volcano from the Volcano House" provides an even clearer illustration. The perceptive visitor should take this foldout to the caldera rim near the west end of today's Volcano House (the foreground of the 1882 scene is now overgrown, and the Volcano House of 1882 is today's Volcano Art Center). The extraordinary scene is rendered even more awesome by the realization that not a square inch of the caldera floor shown in this 1882 illustration is visible today. The small areas of volcanic activity observed by Dutton soon overflowed, covering nearly the entire floor of the caldera by 1885. Isolated among still younger flows, a small area of 1790 lava persists near the Halema'uma'u parking area but is not visible from this viewpoint.

With perhaps one or two exceptions, the pre-1882 history of Kīlauea's caldera is much as Dutton proposed in Chapter IV. But the floor at its far (south) end now is some 300 feet higher than it was in 1882. Here the caldera has overflowed twice since Dutton's visit. The black mounds and nearby flats prominent in the middle distance of the foldout are long gone. The lava lake that they marked was soon replaced by a wide inner pit that perpetuates the name of the lake: Halema'uma'u. On the morning of November 28, 1919, the surface of this lava lake quietly dropped about 600 feet. It remained at that level for a few days, producing an impressive pit 1,200 feet wide before returning to something like its earlier level. In 1924 a huge explosive eruption suddenly restored the pit and gave it the present width (3,500 feet). Large fragments of the island's basement rocks can be seen nearby and blasted far and wide.

Explosive eruptions had occurred here in the past, but at such long intervals that no geologist had noticed. Local *kama'āina* told Dutton about a famous 1790 eruption that destroyed one wing of a Hilo chief's army and demoralized the rest, leading to a decisive victory by Kamehameha I. Bones of victims were visible well into the twentieth century, but Dutton had come to have a very low opinion of the local people, and he dismissed their report as a tall tale told by credulous "savages"—a derogatory term he applied only to inhabitants of the caldera region—and thus missed a "discovery" of the first magnitude. (It probably served him right, but indignant readers should remember that these *kama'āina* inhabited what then was the most remote part of the Big Island, with little opportunity to learn the social skills of their compatriots elsewhere.) When James Dana returned to Hawai'i in 1887, he had no difficulty identifying ejecta from the 1790 eruption.

The remainder of Chapter IV is especially notable for Dutton's masterful use of color to paint his narratives: "the ruddy gleams of boiling lava. . . ," "black as midnight . . . ," and the like. True, occasionally he fell back upon the "pathetic fallacy" beloved of many Victorian writers ("Desolation and horror reign supreme . . .") but this merely strengthened his word pictures of beauty and the seething excitement of volcanic energy displayed for all to see. It is truly tragic that Plates VIII and IX ("The new lava lake, Kilauea" and "Halemaumau—the great lava lake") could not reproduce the rich color of the scenes; viewers would have been astonished. Like the illustrations in Dutton's Grand Canyon masterpiece, these were by artist-cum-geologist William Henry Holmes, whose illustrations provided great insight into the glories of the canyon. But those were reproduced in glorious color. Many puzzled readers of both volumes must have wondered what was so spell-binding about boiling lava.

Dutton extended his reporting to associated volcanic features, particularly the huge "pit craters" east of the caldera. Study of these strange rimless cavities still is at the cutting edge of volcanology. As Wilkes and Dana noticed 40 years earlier, these wide pits characteristically show no evidence of explosive eruptions nor of overflows of lava. Like the old Halema'uma'u pit, they are features of subsidence of lava lakes, but several mechanisms can produce them. A different type of pit occurs along the celebrated Great Crack, actually a series of near-parallel crevices in Kīlauea's southwest rift zone near the overlapping boundary between Kīlauea and Mauna Loa volcanoes. In Dutton's day, this was called "16 (or 17) Mile Crack." Today it is especially noted for pits and crevice caves, one of which has been mapped to a depth of 600 feet. It and a nearby pit crater helped conduct Kīlauea's molten lava toward the sea in deep subterranean channels.

Purely by observation of their lavas, Dutton correctly noted that Kīlauea and Mauna Loa are separate volcanoes. Today we know that the chemistries of their lavas are quite distinct. But now we also know that their eruptions are slightly reciprocal. Mauna Loa normally has an eruptive cycle of about 5 years, but during the present 20-year eruption of Kīlauea, it scarcely has erupted once. Thus there must be some indirect pressure relationship deep in the Earth. Dutton also noted "three classes" of volcanic fumes emitted in and around the caldera. In theory, a single body of magma underlies the entire area, and the different chemistries of the fumes are attributed to different interactions with water and other chemicals deep in the rock. The situation remains enigmatic, however, and much remains to be learned at Kīlauea caldera.

Downslope from the pit craters, Dutton sweeps the reader down the old Chain of Craters Road to the northeastern tip of the island and onward to Hilo. Alongside the ocean, he passed through the now-buried villages of Kealakomo, Waha'ula and Kalapana, then along what now is Highway 137—"the Red Road"—to 'Opihikao, Pohoiki, and Kapoho. Then he turned northwest past Nānāwale on the Puna Trail. A great strength of this chapter is its glowing account of the *kama'āina* of Puna, in dramatic contrast to the so-called "savages" of the caldera region. The people of Puna, Dutton wrote, "are amiable, hospitable and peaceful to the last degree." Universally literate in the Hawaiian language,

> they correspond most vigorously, and the mail facilities are remarkably good. . . . The [weekly postboy's] saddlebags are full of letters and weekly newspapers from Honolulu, printed in the Hawaiian language. This does not seem very barbaric. . . . The native of Puna enjoys the security of his property, his life, rights and liberties by the same title as the American citizen, and in equal measure.

Dutton on Mauna Loa: Chapter V

This chapter perhaps is Dutton at his literary best: "Ainapo is a charming spot in the summertime . . . among open grassy parks and groves of koa [trees]. The air is cool and the trade-wind ever blows gently." It remains so today, although giant eucalyptus trees have replaced many of the *koa*. Elsewhere the chapter seems very modern in discussing the various regions through which his little party traveled. Except for the summit caldera, all are little changed today. Equally modern were some of the concepts he espoused here: sinking of volcano summits to form calderas, and the importance of cave-sized lava tubes in distributing fluid lava great distances on Mauna Loa and Kīlauea.

At the rim of Kīlauea, Dutton had ascended 4,000 of Mauna Loa's 13,677 feet. But he could not carry enough supplies and had to make a round trip to Wai'ōhinu at this point. Along the way, he employed the services of a reliable guide who knew the ancient 'Ainapō summit trail. From a comfortable campsite at timberline, just 7 miles from what then was the 'Ainapō Ranch, "it was only a 5 hour ride to the summit." Later he admitted that they had to lead the mules about halfway. Even so, the ascent was vastly different from the earlier struggles of Wilkes' over-large party, led astray by a boastful guide who only pretended to know the trail.

But Dutton had some difficult work to accomplish before he struck out for the summit, and his 10 days high on Mauna Loa proved to be a considerable test of human endurance. Initially he sought the source vents of the brand-new 1880–1881 lava flows, which reached all the way to what now is the University of Hawai'i's Hilo campus. The uppermost vent is high on the eastern ridge of Mauna Loa, just 1,800 feet below the summit and separated from the summit trail by wide strips of jagged *ʻaʻā* lava. "The fetlocks of the mules are lacerated as they flounder through it. The poor beasts dread the ordeal not a little. . . . " With typical self-effacement he downplayed the inability of the mules to make the last mile to the vent and his struggle on foot through the endless jumble of sharp, clinkery *ʻaʻā*. As he ascended he waxed poetic: "The best conception of [Mauna Loa's] magnitude is to be obtained by attempting to traverse any limited district of it on foot. Mile after mile may be traversed but the landscape seems ever the same. All the great landmarks seem to stand just where they stood an hour before."

Although it is not clear in this chapter, Dutton later recounted having ascended twice from base camp to Mauna Loa's summit caldera, "from two different lines of approach."[8] While inactive, its caldera was so much larger than that of Kīlauea that Dutton was mightily impressed. He would see it differently today; subsequent summit eruptions have largely filled it with placid-looking *pāhoehoe*. In 1882 it was "a wilderness of seething currents of [solidified] pahoehoe" just two years old. "It is all as fresh as if it had solidified only yesterday. . . . The strongest feeling impressed on the mind was that of superlative calm, solitude and desolation." These sentiments remain true today. Although hardly more than a 5-hour hike from the NOAA Observatory on the north side of the volcano, it remains rare to see another human on the summit tableland and caldera.

In this and other chapters, Dutton also recorded keen observations of weather conditions and patterns. Those affected by what we today call vog may take some scant comfort in how much worse it was when Kīlauea was in full eruption—at least in Kona, lowland Maui, and Oʻahu: "At sea level the atmosphere is always hazy, whether over the ocean or over the land, so much that it is rarely possible to see one island from another. . . . It is impossible to obtain a [photographic] picture of details at a distance exceeding one mile." The delicious clarity of the atmosphere above trade-wind level (and in rainy Hilo and Hāmākua) was and is no small matter for rejoicing.

Dutton and Mauna Kea and West Hawaiʻi: Chapters VIII through X.

One of the few literary problems with "Hawaiian Volcanoes" is that it peaked too soon, with Kīlauea and Mauna Loa. Dutton later wrote to James Dana that "I spent a great deal of time in the study of Mauna Kea."[9] Perhaps because he already had written in such glowing detail about generally similar lavas on Kīlauea and Mauna Loa, however, his account of 13,796-foot Mauna Kea seems disappointingly scant. Or perhaps he was

[8] Dutton's letter to Dana, 1883.

[9] Ibid.

merely bone-tired, and windward Hawai'i clearly was not his favorite region. Not far above Hilo, 4 miles of progress required as much as 6 hours in the general area of what is now Kaūmana, at the start of the Saddle Road. "The trail is a mixture of roots, mire, and fragments of rotten fern trees.... Every few rods some poor animal sinks his forelegs or hindlegs into rough, pasty mud and must be unloaded and pried out." Yet when he reverted to his field geologist persona, his accounts were little short of brilliant. Describing the geological differences between the two sides of the island, and the contrasting faces of smooth-looking Mauna Loa and pockmarked Mauna Kea, he writes: "Glancing back at Mauna Loa, not a cinder cone of normal type is anywhere visible upon all its mighty expanse, but Mauna Kea consists largely of fragmental material...."

Dutton spent several unproductive days in the "saddle" between Mauna Loa and Mauna Kea. Those interested in retracing his steps will find his second camp just over a mile east of the junction of the Mauna Kea summit road with Highway 200—the Saddle Road. Clearly his observations on Mauna Kea were almost as acute as on Mauna Loa and Kīlauea. On the main slopes of the mountain he wrote mostly about the cinder cones that pockmark its summit and eastern flank, and the erosion that began its work simultaneous with their fiery origin. Even with no chemical or microscopic analysis he correctly differentiated the two types of basaltic lava on the mountain, and he recorded the ancient adze quarry near its summit. Perhaps because he had ascended to some 13,000 feet four times in 10 days, he failed to mention the geological evidence of glaciation at the summit, the presence of its neat little lake, and the surprising absence of cavernous lava tubes along his ascent route. In contrast, when James Dana returned to Hawai'i to write *Characteristics of Volcanoes* he did not climb Mauna Kea at all.

Back at lower elevations, Dutton followed the general course of Highway 200 through what is now the U.S. Army's Pōhakuloa Training Area. Invigorated by a stream of "delicious water" from Kohala volcano's summit peat bogs, he gladly found the "little village" of Waimea a welcome oasis. He ventured briefly toward the Hāmākua Coast, but it was too wet and too boggy and dissected by too many steep ravines to enchant him. He turned downhill toward Kawaihae and the bone-dry Mauna Loa lavas to the south. En route, he paused to document King Kamehameha's great Pu'ukoholā Heiau in a notably historic photo reproduced in his report as an engraving (Plate XXII). Along what is now the Kohala resort coast, he made good use of old stone-paved trails heading south. At Kīholo Bay the team turned diagonally uphill, across the northwest rift zone of Hualālai volcano. Here they camped somewhere above the neat little volcanic shield of Puhi-a-Pele.

Dutton wrote comparatively little about Hualālai volcano, and it is not clear that he climbed above what is now the Makalei Hawai'i Country Club. He mentioned the shaftlike open vertical volcanic conduits that differentiate Hualālai from all other volcanoes in Hawai'i: "chimneys open with sharp edges at the mouths of the pipes." But he could have observed these at Puhi-a-Pele. He erred by 10 years in specifying the date of its last eruption. Forty years earlier, Dana got it right: 1801.[10]

Continuing south, Dutton said nothing about Kailua-Kona (then a significant center of culture) nor of Pu'uhonua o Hōnaunau, now a national historical park. Evidently he stayed on the upland trail now followed by highways 180 and 11: the Hōlualoa Road. Clearly, however, he had regained his vigor and his sense of awe and rapture: "For three days our journey lay through a country where every turn of the road opened visions of paradise." He was intrigued by the dense field system of agriculture and the numerous ruins: "The general aspect is that of a country once cultivated but long since left to solitude and overrun with untamed vegetation." He anticipated revival of the coffee industry (then in disarray), and reported that pineapples and bananas had reverted to the wild. At Kealakekua Bay he described the great Pali Kapu o Keōua, but he failed to note that it is a fault slip and overlooked its famous burial caves near the present-day Captain Cook memorial.

Approaching Ka'ū District, the overgrown trail changed to a rock-paved highway of sorts, built in a straight line across huge beds of *'a'ā* lava. It led to the famous Kahuku Ranch, now largely a part of Hawai'i Volcanoes National Park. He found the ranch a green oasis, "surrounded by black fields of lava, now only 15 years old"—the 1868 flows of Mauna Loa. Succinctly he discussed these eruptions and the extraordinary series of earthquakes that accompanied them. Then, from Kahuku only 7 miles remained to complete his great loop around the island.

Unwisely, Dutton chose to conclude his chapters on Hawai'i Island with a lengthy technical dissertation on "The Volcano Problem"—Chapter XI. But the knowledge necessary to discuss "the volcano problem" intelligently did not yet exist. This chapter satisfied no one, least of all Dutton. He continued to mull over the problem all his life.[11] Chapter XI is omitted from this reprint.

Dutton on Maui and O'ahu: Chapters XII and XIII

Dutton had spent months visiting only one island and must have been short of time. He returned to O'ahu, then caught the little interisland steamer to Mā'alaea Bay on the south side of Maui Island. He acknowledged that "My visit to Maui was far too brief for my own satisfaction. . . . But to have visited all points of interest would have required many

[10] Cf. note 7 above.

[11] Dutton's last scientific contribution was "Volcanoes and Radioactivity" in 1908 (see list of Dutton's principal writings).

months!" Actually, his tour was considerable. From the sea, he observed the west and south flanks of the old West Maui volcano and the entire south flank of Haleakalā. He ascended the latter via Ha'ikū and Olinda. "Of all places that I have seen or read of none approaches more nearly to my conception of Paradise."[12] At the rim of its summit crater,[13] his literary eloquence again took a Victorian turn to contrast with his paeans of beauty: "A solemn stillness and an air of superlative desolation brood over the scene. Every detail and lineament of volcanic action looks as recent and as fresh as if the fires were quenched and the thunder of the eruptions had ceased yesterday."[14]

Actually he concluded that its last eruption "took place only a very few centuries ago"—not at all a bad guess considering that he never came near Haleakalā's dying gasp, the little Kalua o Lapa flow far down its southwest flank, then not even a century old.

From the summit crater, Dutton accomplished a true tour de force. In a single day he first led his little party down the north face of the huge mountain, a route few travel today. This required descending 7,600 feet through the rough lava of the Ko'olau Gap, then hiking and riding another 20 miles, up and down jungly, rain-slicked erosional gorges along what is now the general route of the Hāna Highway, long after nightfall. "Men were weary and brutes well-nigh used up." The next day may have been worse. Returning from Hāna toward the isthmus to catch the steamer, men and beasts simply gave out in the gorges. Dutton had to hire a small native craft for what must have been a hair-raising night voyage "in the midst of a drenching rain." It did not dampen his enthusiasm for Maui: "magnificent gorges," "a little railway of very narrow gauge and traversed by baby locomotives," "the vast dome of Haleakala. . . . the crescentic shoreline of Kahalui [*sic*] Bay, curving off its beach of white coral sand in exquisite perspective. Nothing is wanting to complete the feature." And perhaps the most significant words in the entire book appear in his half-sentence about "the triumph of energy and enterprise" that created Claus Spreckels' huge sugar plantation. No matter what today's reader thinks (or Dutton

[12] Not all the island of Maui qualified as Paradise. Infamous Big Flea Cave is indicated (but not named) alongside the Haleakalā Trail on the map entitled "Caldera of Haleakala." Evidently Dutton's little party did not have to pitch camp in its less-than-inviting shelter.

[13] The jury still is out as to whether the great sunken area at the summit of Haleakalā is a compound crater or a caldera. Dutton considered it a caldera, but modern geological thought is tilting in the other direction.

[14] This phrase appears to be a figure of speech. The only thunder accompanying the famous May 18, 1980, eruption of Mount St. Helens in Washington resulted from incidental lightning discharges. A U.S. Forest Service employee at Red Rock Pass, high on the southwest side of that volcano, heard nothing during the eruption. He was unaware of it until he glanced upward and saw a great ash cloud mushrooming above him.

himself thought) about Spreckels' business ethics, sharp American businessmen could (and did) grow rich under the Hawaiian status quo. Ergo, annexation was unwarranted.

Although Dutton later wrote Dana that he "went over the island of Oahu pretty thoroughly,"[15] his final chapter is brief. Nowhere is there mention of the people of the island, king or commoner, *kama'āina* or *haole* (Whites). Most is a somewhat stilted discourse on erosion, perhaps necessary to defend previously published statements about the process in arid terrains. The delightful engravings of an unpopulated Waikīkī (labeled "Diamond Head") and of Nu'uanu Pali ably replace pages of text.

Yet Dutton's O'ahu observations were no less acute than on the other islands: the pair of volcanoes that formed the Ko'olau and Wai'anae ranges; late resumption of volcanic activity after the island had moved off the hot spot; the predominantly windward erosion; reef limestone up to 200 feet above present sea level; and much more. Many decades in the future were concepts such as the stupendous size of the Kailua landslip that formed the "gigantic windward cliffs." But he at least alluded to the origin and development of the famous volcanic landmarks of O'ahu: the Pali, Diamond Head (whose name had stabilized since the Wilkes Expedition called it "Diamond Hill"), the Punchbowl, Hanauma Bay, the windward islets, and others. His spirited prose never faltered: "The traveller who visits these islands, even though it be for the few hours allowed by the stoppage of the Australian steamer as it touches at the port, will most probably take a carriage to visit the Pali, as it is termed. Here there suddenly breaks upon him one of the most beautiful and picturesque views in the world."

But something important clearly is missing in this chapter: it is curiously truncated, without a summing-up of why "Hawaiian Volcanoes" was important to its readers. Would an accomplished author deliberately end so important a scientific work so abruptly? Was a strongly worded concluding paragraph hastily deleted as the manuscript went to press? And if so, by whom and why? This matter is discussed further in Appendix A, but the answers may have been lost forever with Dutton's trunkful of personal papers.

William R. Halliday
Hilo, 2004

[15] Dutton's letter to Dana, 1883.

Illustrations

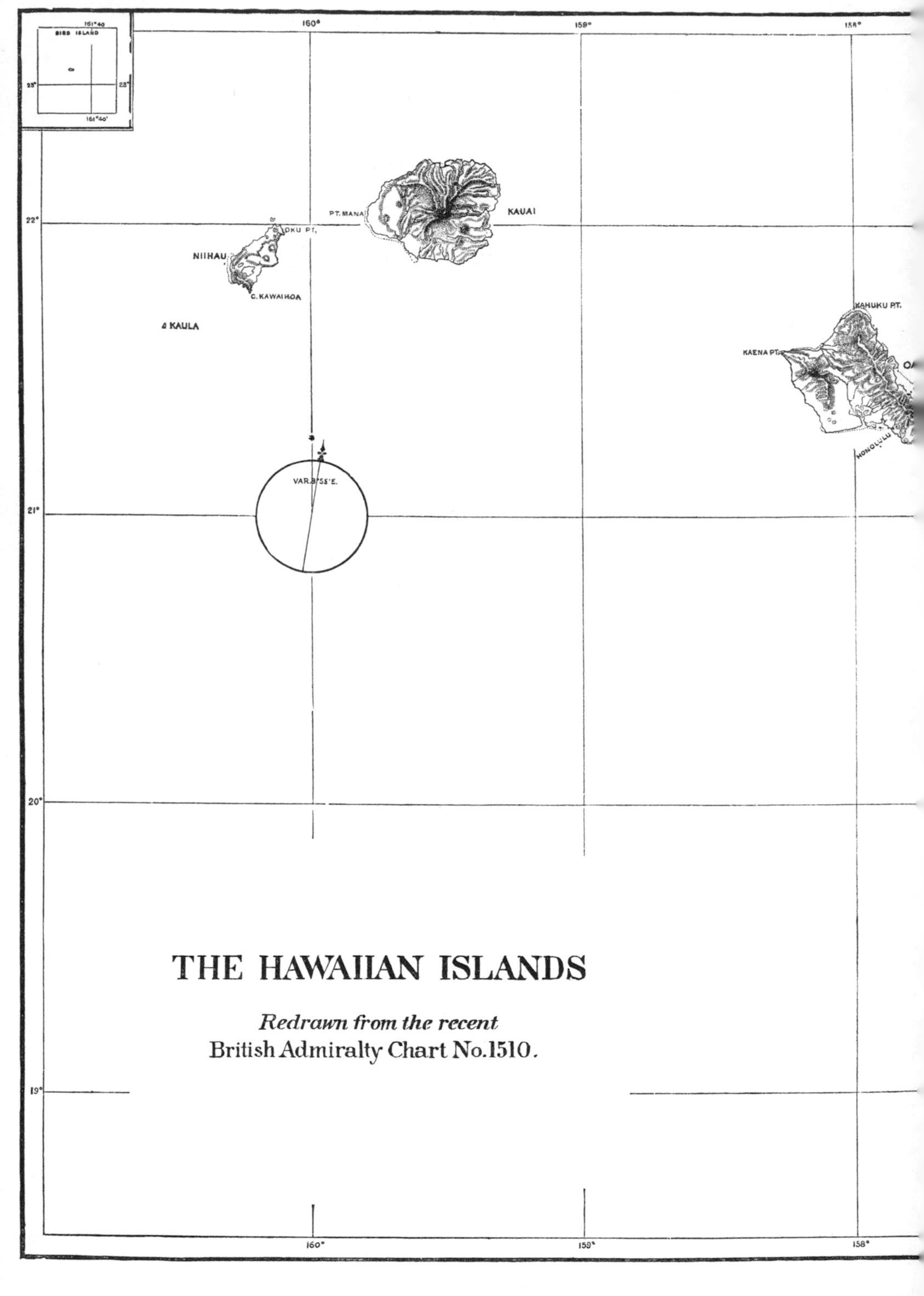
BIRD ISLAND
161°40'
23°
160°
159°
158°
22°
21°
20°
19°
PT. MANA
KAUAI
OKU PT.
NIIHAU
C. KAWAIHOA
KAULA
KAHUKU PT.
KAENA PT.
HONOLULU
VAR. 8°55' E.
THE HAWAIIAN ISLANDS
Redrawn from the recent
British Admiralty Chart No.1510.

157°
156°
155°
22°
21°
20°
19°
MOLOKAI
MAUI
LANAI
KAHOOLAWE
KAMALALAEA B.
KANIKI HD.
UPOLU PT.
HAWAII
C. KUMUKAHI
KA LAE

HAWAIIAN VOLCANOES.

BY CAPT. C. E. DUTTON.

CHAPTER I.

GEOGRAPHY OF THE HAWAIIAN ISLANDS.

The Hawaiian Islands are situated between the meridians 154° 30′ and 160° 30′ west of Greenwich and between the parallels 18° 40′ and 22° 15′ north. A line of the least curvature drawn through the axis of the group, or through the centers of gravity of the areas of the principal islands, would approximate roughly the segment of a large circle convex towards the northeast. The chord of this segment connecting the most widely separated points of the extreme islands would have a course about north 60° west, and would have a length of about 350 nautical miles, or 400 statute miles. The distance of the group from San Francisco may be stated at 2,000 miles; the steamers of the Pacific Mail Steamship Company, which sail monthly to Australia, reckon the distance to Honolulu at 2,080 miles, and make the trip to that port in about seven days.

Twelve islands are usually spoken of as composing the group. Four of these are mere barren rocks which may count for nothing. Of the remaining eight, four are large islands and four are of much inferior though still notable extent, and quite habitable. The following list shows the approximate extent of the habitable islands:

Names.	Length.	Breadth.	Areas.
	Miles.	*Miles.*	*Sq. miles.*
Hawaii	90	74	3,950
Maui	48	30	620
Oahu	46	25	530
Kauai	25	22	590
Molokai	40	7	190
Lanai	17	9	100
Niihau	20	7	90
Kahoolawe	11	8	60

All of these islands are volcanic. No other rocks than volcanic are found upon any of them excepting a few remnants of raised sea beaches composed of consolidated coral sands. All the larger ones are very mountainous and lofty. The culminating points of the island Hawaii are Mauna Kea, 13,900 feet, and Mauna Loa, 13,700 feet high. The summit of Háleákalá on East Maui is 10,350 feet, and the mountains of West Maui attain 5,900 feet.* Oahu has peaks on the eastern side of the island 2,900 feet high, and others on the western side 3,850 feet high. The loftiest point of Kauai is 6,200 (?) feet above the sea. All of these data are referred to sea-level. Deep-sea soundings in the vicinity have recently disclosed the fact that these volcanic piles are only the summits of gigantic mountain masses rising suddenly from the bottom of the Pacific, which for many hundreds of miles around them is only moderately diversified. The slopes of Mauna Loa east, west, and south descend beneath the surface of the ocean with a gradient fully equal to, if not greater than, the visible slope of its flanks. The submarine slopes of all the other islands in directions perpendicular to the principal

---

* Since Hawaiian names must often occur in the following text it may be desirable to indicate a few simple rules for their pronunciation.

1. The consonants all have the German sound except w, which is pronounced as in English; or better still, they all have the English sound without exception.

2. The simple vowels generally have the pure German sound. Thus *a* sounds as the German *a* in *fahren; e* has always the same sound as *e* in *mehr; i* is the same as the French *i*, or as the German *ie*. The *o* is the same as in all European languages. The *u* has the sound of *oo* in English or *u* (without the umlaut) in German. There is no other *u*-sound than this in the language.

3. Vowels in juxtaposition often have the effect of diphthongs. Thus *au* has the German sound as in *haus; ao* together give the same sound as *au; ai* has the sound of the German *ei;* and *ae* sometimes occurs with the sound of the German *ei*. I believe there are no other vowel combinations which produce diphthong effects. Hawaiian scholars deny that there are any real diphthongs in the language and maintain that there are only five pure vowel sounds, *a, e, i, o, u*. The apparent diphthongs are merely the inevitable result of the rapid enunciation of two consecutive vowels. Thus if anybody will take the sounds of *a* and *u* or *a* and *o*, as above described, and repeat them over and over again as rapidly as possible he will get the German *au* sound in spite of himself. So with *a* and *i*, he will get the German *ei*. The rapid pronunciation of *a* and *e* will also give the German *ei*. The only virtual diphthongs are *au, ao, ai, ae*.

4. There are no silent letters in the written Hawaiian language.

5. No invariable or easily-applied rules for the accentuation of syllables can be given, but the following will cover a large majority of cases: (1) Accent every alternate vowel or diphthong. (2) Arrange the alternate accents in such a way that the final syllable (which is always a vowel or virtual diphthong) shall be unaccented. Thus, Ho′-no-lu′-lu, Lau′-pa-ho′-e-ho′-e, O-a′-hu, Mo-ku′-a-we′-o-we′-o, Ka-ho′-o-la′-we. Perhaps the most important exceptions to this majority rule of accentuation which we shall encounter in the following text will be in the name Ha′-le-a′-ka-la′, where the accent falls on the last syllable; and in the name Ka-u′, where not only is the accent pressed strongly upon the last vowel but the *a* and *u* sounds are kept as distinct as possible and not permitted to run together into a virtual diphthong. Where a word ends in two vowels which tend to run together into a diphthong, the virtual diphthong so produced is almost sure to be accented. Thus Mo′-lo-ka′-i would conform to the majority rule if the last two vowels were kept perfectly distinct. But in practice they are not; so that the word becomes Mo′-lo-ka′i. Perhaps a large class of exceptions can be made to conform to the rule in the following way: When a word ends in a virtual diphthong resolve the terminal diphthong (but not the preceding diphthongs) into its two constituent vowels and then apply the alternate accents so that the last resolved vowel shall be unaccented.

axis of the group are equally great and possibly somewhat greater. The depths attained by these continuous slopes within 30 to 50 miles of the shores vary from 2,400 to 3,100 fathoms, or 14,000 to 19,000 feet. Mauna Loa and Mauna Kea, referred to their true bases at the bottom of the Pacific, are therefore mountains not far from 30,000 feet in height. And in general the island group consists of the summits of a gigantic submarine mountain chain, projecting its loftier peaks and domes above the water.

There is reason to believe that this same submarine chain continues in great force many hundred miles to the west-northwestward of Kauai, for the charts of that portion of the Pacific show in that direction minute islands and shoals strung along at intervals of fifty to a hundred miles. These are all in the continuation of the main axis of the Hawaiian group. In the year 1875 the Challenger expedition ran a line of soundings from Japan to the Hawaiian Islands, about 300 to 500 miles south of this presumptive range and parallel to its trend. A remarkably uniform depth of 2,500 to 3,100 fathoms was found.

The magnitude and proportions of this great mountain range may be illustrated by a diagram drawn to scale, representing in the most general way its transverse profile (Pl. III). The submarine portion of this profile must be interpreted with the greatest latitude. The soundings have been made at intervals too wide apart and along lines much too few to give us any knowledge of its details. We do not know whether it descends with a uniform sweep or goes tumbling down in ragged, ridgy confusion. Further knowledge of the configuration of the sea-bottom in the vicinity of these islands is greatly to be desired.

On the island of Hawaii the volcanic forces are still in operation. On the eastern portion of Maui, they have rested at a very recent epoch. In the other islands they have long been extinct, and the piles they built up have been greatly ravaged by erosion. On Hawaii there are at present two grand foci of volcanic eruption where the fires are still raging. These are Mauna Loa and Kilauea. Just west of Mauna Loa is the large volcano Hualalai, which gave forth three eruptions in the early part of the present century, but which has been dormant since 1811. Mauna Kea gives evidence of having reposed for many centuries—or throughout a period which, historically considered, may have lasted some thousands of years; though, if we value time by the geological scale, the date of its last activity would be regarded as very recent. Haleakala on Maui gives indications of considerably greater recency in its last eruptions than Mauna Kea, but the natives have no traditions of any outbreak from it, and we may infer that it has been quiet for several hundred years. With regard to the other volcanic centers, we have no means of judging of the antiquity of their final action except the progress made by erosion in demolishing them, and this progress is, in every instance, considerable. It is most conspicuous on Kauai and Oahu, and almost equally so on Molokai and West Maui. From this it is inferred that the western islands of the group have longest enjoyed immunity from eruption. Kauai, especially, is frequently spoken of as the oldest island of the group, and judging from the amount of destruction wrought upon it by the eroding forces the statement is in some measure apparently justified—but only to this extent: the period which has elapsed since the cessation of eruption has probably been longer there than in the other islands. It does not follow, however, that the eruptions began there any earlier than on Hawaii. Whether they did so or not is a question which I see no way of determining.

Mauna Loa—"The Great Mountain"—is certainly the king of modern volcanoes. No other in the world approaches it in the vastness of its mass or in the magnitude of its eruptive activity. There are many volcanic peaks higher in air, but they are usually planted upon elevated platforms, where they appear as mere cones of greater or less size. Regarding the platforms on which they stand as their true bases, the cones themselves and all the lavas which have emanated from them never approach the magnitude of Mauna Loa. Ætna and all its adjuncts are far inferior, while Shasta, Hood, and Ranier, if they were melted down and run together, would fall much below the volume of the island volcano. We do not know at what level the base of Mauna Loa is situated. We only know that it is below sea-level, and probably far below it. But on the other hand, it may not be so low as the adjoining depths of the Pacific, for, as will appear in subsequent chapters, there is evidence that its platform has been hoisted, and to a considerable amount, during the progress of its eruptions.

Mauna Kea—"The White Mountain"—is also a colossus among volcanoes, and, in truth, I do not recall another in the world which is equal to it in magnitude, except its neighbor, though possibly Mount St. Elias may be nearly so. The summit of Mauna Kea is a trifle higher than that of Mauna Loa, but its slopes are steeper, and its base is therefore much smaller. The magnitude of Mauna Loa is due chiefly to the great area of its base, which is nearly elliptical in shape, with a major diameter of 74 miles and a minor of 53 miles, measured at sea-level.

In the aggregate of its eruptions Mauna Loa is also unrivaled. Some of the volcanoes of Iceland have been known to disgorge at a single outbreak masses of lava fully equal to them. But in that island such extravasations are infrequent, and a century has elapsed since any of such magnitude have been emitted, though several of minor extent have been outpoured. The eruptions of Mauna Loa are all of great volume, and occur irregularly, with an average interval of about eight years. Taking the total quantity of material disgorged during the past century, no other volcano is at all comparable to it. A moderate eruption of Mauna Loa represents more material than Vesuvius has emitted since the days of Pompeii. The great flow of 1855 would nearly have built Vesuvius, and those of 1859 and 1881 are not greatly inferior.

Mauna Loa and Kilauea are in many important respects abnormal volcanoes. Most notable is the singularly quiet character of their eruptions. Rarely are these portentous events attended by any of that extremely explosive action which is characteristic of nearly all other volcanoes. In only one or two instances within the historic period, have they been accompanied by earthquakes or subterranean rumblings. The vast jets of steam blown miles high, hurling stones, cinders, and lapilli far and wide, filling the heavens with vapor and smoke, and hailing down fragments and ashes over the surrounding regions, have never been observed here. Some action of this sort is indeed represented, but only in a feeble way. The lava wells forth like water from a hot, bubbling spring, but so mild are the explosive forces that the observer may stand to the windward of the grandest eruption, and so near the source that the heat will make the face tingle, yet without danger. Ordinarily the outbreak takes place without warning and without the knowledge of the inhabitants, who first become aware of it at nightfall, when the sky is aglow and the fiery fountains are seen playing. As the news spreads hundreds of people flock to it to witness the sublime spectacle, and display almost as much eagerness to approach the scene of an eruption as the people of other countries show to get away from one.

A direct consequence of this comparatively mild and gentle behavior is the absence of those fragmental products which form so large a proportion of the products of other volcanoes. The acute cones which terminate the summits of the volcanic masses of the Mediterranean, the Andes, the Cascades, the Philippines, are composed very largely of scoria, cinders, ashes, and lapilli. Around them and on their flanks may be seen a tumultuous throng of parasitic cinder-cones formed of lapilli blown upward and showered down around orifices from which lavas have in most cases been extruded. On Mauna Loa and Kilauea few piles of such fragmental matter are seen, and such as occur are generally insignificant in size and abnormal in appearance. On neither of them is there any summit cone. Their tops are broad, flat tables, with immense pits sunken in them. The ejecta are altogether massive lavas and almost nothing else. This is all the more noteworthy, because their giant neighbors, Hualalai and Mauna Kea, as well as Kohala Mountain, at the northern extremity of the island, are thickly covered with cinder cones of most typical form and structure. Neither of the three last mentioned, however, have any large dominant terminal cone overpowering all the rest, like Ætna, Teneriffe, and Shasta, but they are sprinkled all over from base to summit with cinder cones which are of identical pattern whether planted high or low. On the broad, tabular summits they appear to cluster more thickly, but they exhibit no marked superiority of size over those below.

The great magnitude of the individual eruptions of Mauna Loa and Kilauea, and the absence of fragmental products, supply an explanation of their abnormally flat profiles. Fragmental ejecta pile up around the

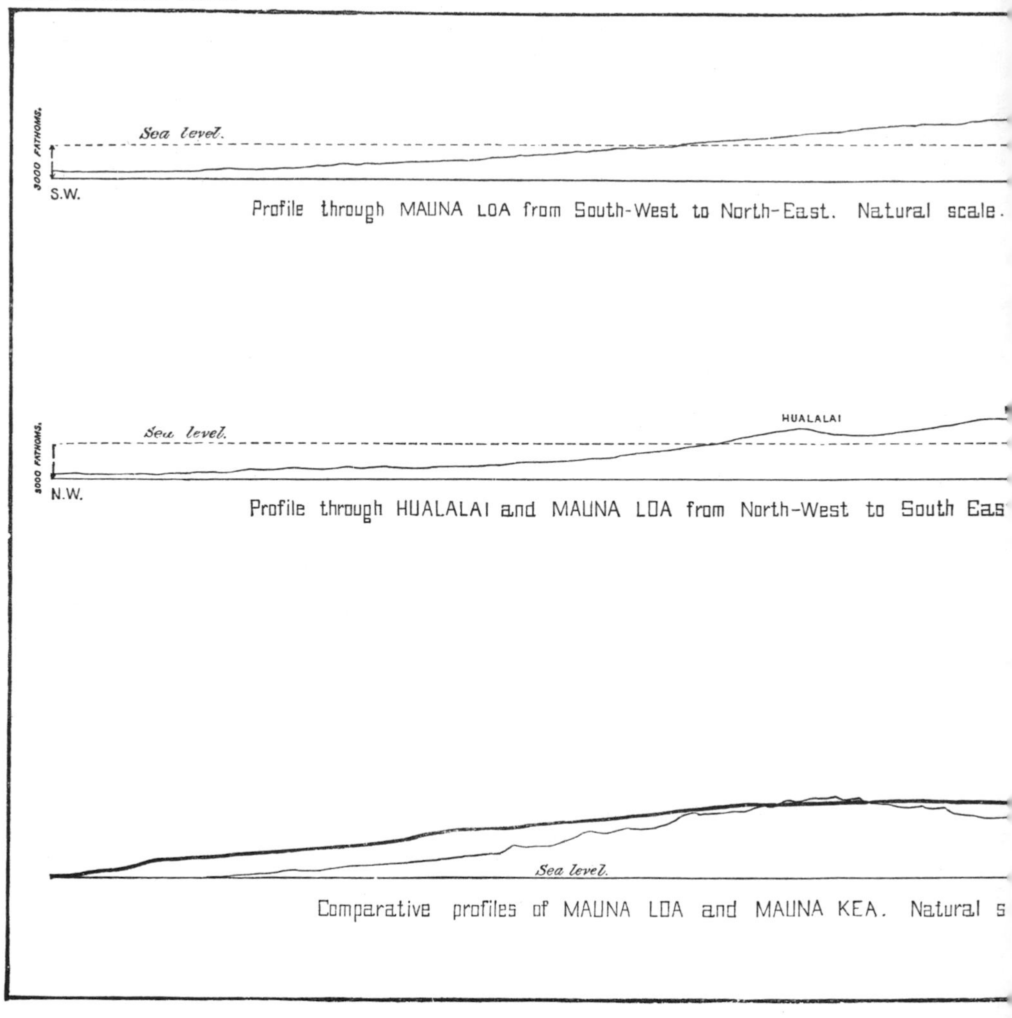

Sea level.
3000 FATHOMS.
S.W.
Profile through MAUNA LOA from South-West to North-East. Natural scale.
HUALALAI
Sea level.
3000 FATHOMS.
N.W.
Profile through HUALALAI and MAUNA LOA from North-West to South Eas
Sea level.
Comparative profiles of MAUNA LOA and MAUNA KEA. Natural s

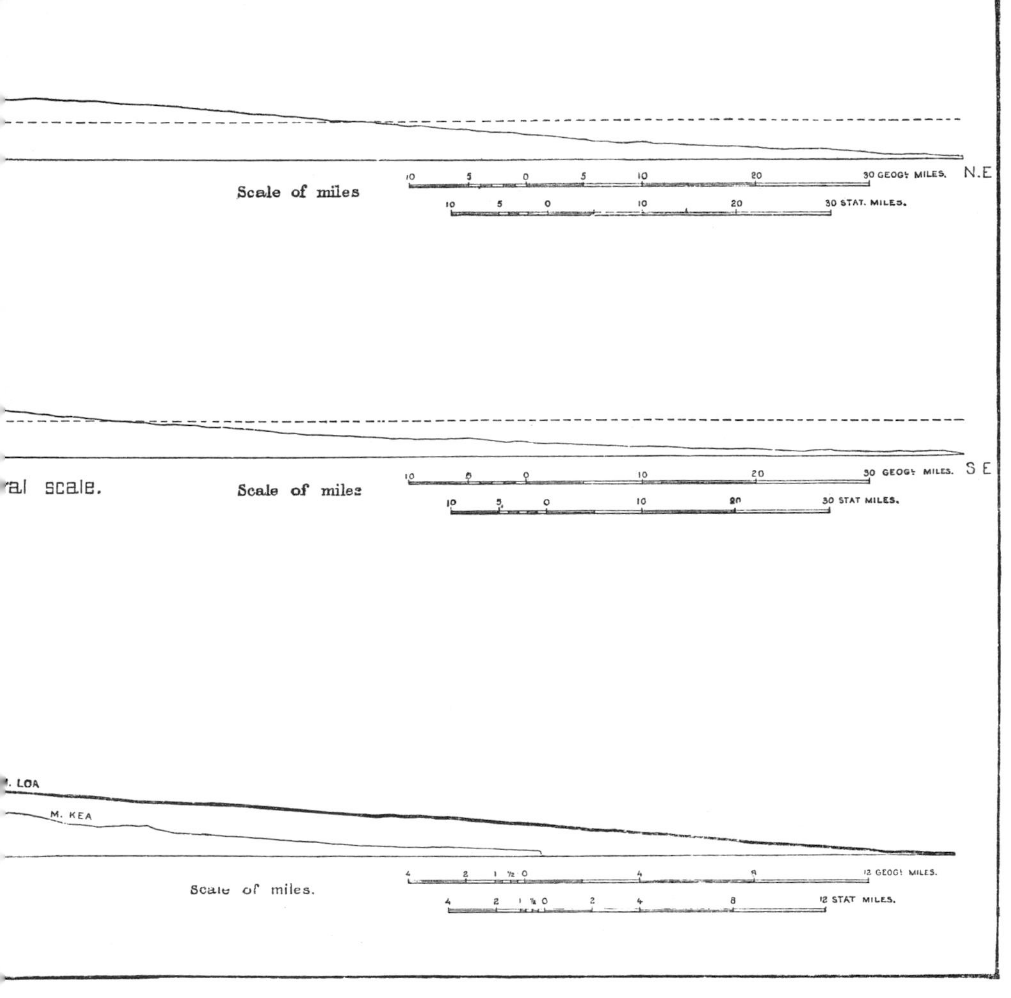

AND MAUNA KEA.

vents from which they are projected; but great streams of fluent lava run far away from their sources.

The large pits or so-called craters which are sunken in the summits of Kilauea and Mauna Loa, and the still more remarkable one on Haleakala, are not entirely unparalleled, though their analogues in other regions are very uncommon. Descriptions of these wonderful features cannot be given here without anticipating too much, and they are therefore reserved for subsequent chapters.

Earthquakes are of common occurrence in the islands, but they usually have their centers of disturbance around Mauna Loa and Kilauea. In the islands to the northwestward the shocks are infrequent and feeble and generally consist of the waves generated at Mauna Loa, and dying out as they recede from it. Those frightful devastating convulsions of the earth which are so calamitous in some other regions sometimes happen in southern Hawaii, but only at intervals of many years. Such a one occurred in 1868. But even at the focus of disturbance the shocks are seldom of a very alarming or destructive character. Small or moderate tremors, however, are very frequent.

The lavas of all the islands are basaltic throughout. Some varieties appear to approach or even correspond to andesite (augitic), but they are not common.* Rocks of the acidic or trachytic type are unknown.

* The general reader is often troubled to understand the meaning of the terms employed in designating the various species of volcanic rocks. It may comfort him to know that his perplexity is shared by the profoundest students of lithology, who find the greatest difficulty in defining the varieties with which they deal. There are, however, some broad and well-settled principles to which it may be well here to advert.

All lavas are complex compounds of silica and alumina, with oxides known as alkaline and earthy bases. Silica is by far the largest constituent in all of them. Second in importance is alumina. Besides these we have two "earths," lime and magnesia, and two alkalis, soda and potash. Iron oxide, too, is rarely absent, and frequently forms a very large ingredient. Thus we have seven oxides which are found in all lavas. Rarely is any one of them absent, though sometimes the quantity of some one of them may be very small. Other oxides are often found; but always under circumstances which lead us to regard them as accidental occurrences.

The proportions of these constituents vary rather widely. The following table shows their ordinary ranges of percentages:

	Per cent.
Silica	50 to 75
Alumina	8 to 18
Lime	2 to 10
Magnesia	1 to 6
Soda	1 to 6
Potash	1 to 6
Iron oxide	1 to 14

Sometimes the percentages are found lying outside of either of the extremes, but not far.

Two primary divisions of lavas have long been recognized. In one division the quantity of silica is very large—say from 62 to 75 or even 80 per cent., and the other constituents are correspondingly diminished. This group has usually been termed the acidic group, from the large quantity of silicic acid (*i. e.*, silica) which it contains.

There is a considerable range in the variety of basalt presented, but the main features of the group are always strongly pronounced. The lavas of Mauna Loa appear to be extremely basic and abound in olivin and augite; indeed the quantity of those minerals, especially olivin, is phenomenally great in most of the recent flows. The composition of the lavas cannot be regarded on the whole as presenting any novel features, though the extreme basicity of many of them is quite striking.

Very pleasing are the studies in the processes and results of erosion which these islands offer to the geologist. In all those portions of the islands where quiet has long prevailed the scenery has habitudes of extreme boldness and animation. It is always picturesque, very often beautiful, and sometimes grand. Cliffs, crags, and cañons are carved in the mountains with as much sharpness and spirited detail as in the Plateau country. In West Maui and Kauai may be found walled valleys and amphitheaters which almost rival Yosemite. The windward fronts of Oahu and Molokai present cliffs 2,000 feet high, carved in a manner which is quite unique and rarely surpassed by the finest sculpture in the valley of the Colorado. On the weather sides of Hawaii and Maui the gentle slopes of the mountains terminate upon the ocean in walls a few hundred feet high, while the platforms are gashed with cañon-valleys which are marvels of beauty. Over all is spread the mantle of a tropical vegetation so rich and splendid that it "makes the pomp of emperors ridiculous."

It is not a little surprising to find those carvings and general details of land sculpture which are produced in the arid climate of our western plateaus again confronting us in the extremely wet climate of these

In the other division the silicic acid ranges from 50 to, say, 62 per cent., and contains a correspondingly larger quantity of alkaline and earthy bases. It is, therefore, termed the basic group. No hard and fast line separates these two primary divisions, but they shade into each other with respect to composition, and even overlap each other. Still the two divisions may be said to hold good just as we may say that an island has an eastern part and a western part, though we may not be able to point out any exact line where the two portions can be sharply and logically separated. The acidic rocks were formerly called trachytes and the basic rocks basalts.

In most cases the two groups are readily distinguished by the experienced eye, but in many instances this immediate distinction is very difficult. As a general rule the traychtes are lighter-colored and of less specific gravity; the basalts are dark or black and heavier. But this diagnosis is too unsafe to be trusted. A more accurate method is found in the study of the minerals or crystals which most lavas contain. This, however, requires an extended knowledge of mineralogy and cannot be discussed here.

The two primary divisions of acidic and basic rocks have proved to be insufficient to meet the wants of lithologists who desire to use great precision in treating of the lavas. This want has led to a subdivision of each division into two groups. The trachytic or acid rocks have by many writers been subdivided into a more acidic and a less acidic portion. For the more acidic group the name rhyolite has been employed, while the name trachyte has been restricted to cover the moderately acidic rocks. Similarly the basalts have been subdivided into a moderately basic group termed andesite, while the name basalt has been restricted to the most basic portion.

islands with such modifications only as seem to be due to the different character of the rocks themselves. Cliffs, cañons, crags, and pinnacles are found here in habitudes as typical as those of Utah and Arizona. The walled amphitheaters of Kauai are as bold and abrupt as those of the Vermilion Cliffs in Utah. The gorges of Maui and Northeastern Hawaii are as truly cañons as many of the tributary valleys of the Colorado, and yet the climates of the two regions offer the strongest possible contrast. In the one region the rainfall rarely exceeds 15 inches per annum, and is, in some localities, less than 8. In the other it ranges from 150 to 240 inches. Many have supposed that these sharp, angular topographic forms in America are due chiefly to the aridity of the climate. But while there can be no doubt that under the circumstances there existing the aridity on the whole favors the production of such features, we have long been aware that it is not to be inferred that they may not be produced without it. The causes which determine the habitudes of topography are very complicated, though capable of being formulated. We now possess a knowledge of them which may not be complete though it is adequate to the solution of most problems, and it is perfectly intelligible that cliffs and cañons may be formed in a moist climate as well as a dry one. Some brief discussion of this will be attempted in subsequent chapters.

The climate of the islands presents considerations of very great interest. It is difficult to find anywhere greater constancy of atmospheric conditions than those prevailing over the ocean within the heart of the trade-wind belt. The temperature varies but little from winter to summer, and the general drift of the atmosphere is rarely interfered with by storms and cyclones. And yet upon the islands themselves it may be said that there are almost as many climates as there are square leagues. And the differences of climatal condition exhibited by localities separated only half a dozen miles are extreme. As a general rule the windward sides are excessively rainy, the precipitation frequently exceeding 200 inches in a year. The leeward sides are generally arid, but to this there are some striking exceptions, especially under the lee of Mauna Loa, where the rainfall is almost as great as on the weather side. The explanation of this apparent anomaly is interesting. In ascending the lofty domes of Mauna Loa we find that above 10,000 feet or even above 8,000 feet the trade wind is no longer felt, however strongly or steadily it may be blowing below. Still higher, at 12,000 to 13,000 feet, we find the drift of the atmosphere to be in the opposite direction. This fact is rendered visible below by watching the drift of the clouds. The trade-wind clouds, which hang low, sail ever to the southwestward while the higher cirrus and stratus float almost in the opposite direction. Wherever the land barrier is low enough to permit the trade wind to blow over it the lee of the barrier is invariably dry, and sometimes is as parched and barren as the sage plains of the Rocky Mountains. The winds throw down their moisture copiously as they rise to the dividing crest, and descend hot and dry. But when the barrier is lofty enough to effectually oppose the drift of the air the lee becomes subject to the simple alternation of daily land and sea breeze. As the sea breeze comes in and ascends the slope it sends down rain. As the land breeze floats downward and outward it is dry and clear. The sea breeze sets in a little before noon, and the land breeze goes out a little before midnight. Relatively to human comfort the climate is perfection. It is never hot, and at moderate altitudes it is never cold. The heat of summer is never sufficient to bring lassitude, and labor out of doors is far more tolerable than in the summer of New England or Minnesota. The air is health itself. Pestilential diseases are unknown, and indeed there is no disease of a local character, or which may be regarded as having any special development in the islands, except syphilis and a closely allied form of leprosy to which the native race is exceptionally susceptible.

Only a small proportion of the area of the islands is capable of sustaining a dense population. The most habitable tracts are near the sea-coast, and only a part, or even a small part, of these are really fertile. The interior portions are mountainous and craggy, with a thin soil, admirable in a few localities for pasturage, but unfit for agriculture. Many parts of the shore-belt are arid and almost barren. Others are covered with lavas too recent to have permitted the formation of soil, and still others are trenched with ravines so deep and abrupt that access is difficult. Deep rich soils at altitudes adapted to the growth of the sugar-cane probably form less than the fortieth part of the entire area. Shallower soils, however, are a little more extensive and yield other crops of tropical staples in abundance. The native food consists largely of the taro plant (*Arum esculentum*), of which the best varieties are grown in shallow ponds of fresh water, and I believe in the correctness of the estimate that about 40 square feet, or say 2 meters square, will yield taro enough to supply one man for a year, this being his principal food. Wet taro lands are limited to a few hundred acres on account of the scarcity of running water; but these few hundred acres would have supported nearly a thousand men per acre! Humboldt's estimate of the food capacity of the banana is lower. Another variety of taro is grow upon the mountain slopes, in moist soil, but it is regarded as somewhat inferior. The sweet potato grows among the rocks in a marvelous way, almost without soil, and flourishes in better ground in a most prolific manner. The common potato sometimes thrives greatly, but is generally ravaged by worms. Wheat was formally cultivated on high lands for the California market, but the islands now receive all their cereal products from California. The coffee plant not only flourishes here, but the quality is equal to that of the most famous and choicest. The "Kona" coffee is superior to the Mocha and equal to the best Liberian. The climate is also very favorable to the growth of the long-staple sea-island cotton, but as this variety must be picked by hand, instead of being ginned, the high price of labor in the islands renders its culture unprofitable.

Tropical fruits of nearly all kinds grow in the greatest abundance, the orange, lemon, lime, mango, pineapple, chirimoya or custard-apple, the alligator pear, pomegranate, and guava, all of which are exotic. The banana is indigenous, and is the most abundant of all fruits. One fruit, very common and peculiar to the Pacific islands, is the ohia-apple, which is soft, juicy, mildly acid, and beautiful to the eye, but rather deficient in flavor. It is about as large as a rather small apple, and contains a single large drupe. Much zeal has been shown by the white residents in introducing many varieties of tropical trees and shrubs, so that now a very large part of the flora is of exotic origin. Many varieties of palms, the choicest trees of India, the caoutchouc, the papaya of New Grenada, the traveler's tree of Madagascar, are among the more conspicuous, and they appear to flourish as well as in their original soils.

The chief industry of the islands is the cultivation of the sugar-cane. For this their soil seems better adapted than any other in the world. The yield will average about 5,000 pounds of sugar to the acre, and choice fields sometimes yield twice that amount. Under the existing reciprocity treaty between the United States and the Hawaiian Islands this industry has expanded with great rapidity until it has brought under cultivation very nearly all of the easily available land which is suited to it. Large amounts of American capital have been invested in the plantations and in the accessory commerce, with great advantage to both countries.

I regret most deeply that I have been unable to procure a good map of the island of Hawaii—the largest of the Hawaiian group and the theater of the great living volcanoes. There is no such map in existence. But fortunately much progress has been made towards the construction of one which will, no doubt, rank high among first-class maps. I am indebted to Prof. W. D. Alexander, the surveyor-general of the Hawaiian Kingdom, for excellent maps of Maui and Oahu, which have been reproduced for this memoir on one-fourth the original scale. The survey now in progress under Professor Alexander is an admirable one; in truth, a model in its way. The circumstances which have led to it are worth reciting.

When the Sandwich Islands were discovered by Captain Cook the people were by no means savages, but had a state of society so well organized as to be quite as much above savagery on the one hand as it was below civilization on the other. In truth, it is with no little surprise that the student finds this social condition to have been remarkably similar to that of Europe in the dark ages. The social system was almost exactly the feudal system. The French and Saxons of the ninth and tenth centuries, no doubt, were greatly in advance of the primitive Hawaiians of the eighteenth century in respect to the arts; but it is difficult to understand in what respect their social organization was any higher. These old Hawaiians held land by about the same sort of tenure as the subjects of Charles Le Gros and King Knut. The king was a petty suzerain in whom alone the allodium vested. The chiefs held lands substantially in fief and for military service, while the common people were mere villeins and tenants at will. Civilization made progress with this people as rapidly as it has with the Japanese. In twenty-five years after the landing of the missionaries (1820) the whole people had, in a great measure, become Americanized. With the great social revolution came the necessity for a change in the system of land tenure from the feudal system to the fee simple, with a record title. Lands had always been subdivided by metes and bounds from time immemorial, and their locations had, in the absence of the art of writing, been handed down—by tradition, indeed, but with an accuracy and rigor for which tradition is only an imperfect expression. To make these titles, as defined in metes and bounds, subjects of court record a survey was necessary. It was undertaken and carried forward on a scale commensurate with its importance, and the Hawaiian Government is to be congratulated in having secured for the work an administrator so able, faithful, and efficient as Professor Alexander. The survey of Oahu and Lanai, so far as the general map is concerned, is complete. Maui is nearly finished. Much has been done on Hawaii, but the end is so far away that some years must elapse before a good map is possible. The survey of Kauai is begun on the new system, but the map of that island now existing is a compilation from former surveys upon a less systematic plan.

I cannot sufficiently thank Professor Alexander for his courtesy in supplying me with copies of maps specially drawn for my purposes. But the want of a good map of Hawaii I lament greatly. So imperfect is the only one at present obtainable that I have felt obliged to refrain from attempting to delineate upon it some very important features connected with recent eruptions, from the fear that they would be misleading.

CHAPTER II.

A JOURNEY TO KILAUEA.

The supreme attractions which these islands present to the geologist are the great volcanoes of the island of Hawaii. This island is very scantily inhabited; is of great extent, and presents certain difficulties in the way of travel which require special preparations to encounter them successfully. Most travelers rely upon the hospitality of the islanders for furnishing the necessary facilities. Although this hospitality is rendered with a heartiness and freedom which no one can recall without emotion, an European or American traveler is generally reluctant to lay himself under such obligations. He soon finds, however, that it is necessary to accept these kindly attentions in the spirit in which they are tendered. But if he contemplates a journey outside of the beaten paths of travel, and desires to see any considerable portion of the island, he must rely upon other resources. I therefore determined to fit out a pack train, after the customary manner of the Geological Survey in its explorations of the Rocky Mountains, and for this purpose purchased at San Francisco and Honolulu the necessary material.

In selecting a point on the island of Hawaii as the base of operations I was governed by the following considerations. Travelers generally land at Hilo for the purpose of visiting the volcanoes. This town is situated upon the rainy side of the island, and wet weather is very obnoxious to the field geologist. The southern side of the island is much drier, and the country more open and free from forest. The great volcanoes are quite as accessible from the south as from the east, and the difficulty of obtaining animals and packers was no greater. I therefore selected the southern or Kau side of the island as the point from which to approach Mauna Loa and Kilauea. A steamer runs weekly from Honolulu to Hilo. A second steamer runs from Honolulu to the southern coast of the island, making the round trip in ten days. Taking passage on the Kau steamer on Tuesday afternoon, I reached the southern side of Hawaii early on the Thursday morning following, landing in boats a little before daybreak. I was most kindly entertained and cared for at the Naalehu sugar plantation, situated near the village of Waiohinu. Here I was so fortunate as to secure nine serviceable animals, and the assistance of a white resident of that village, named Macomber, who had resided on the island for about thirty years. He was a clever tradesman, being a good carpenter, blacksmith, and saddler, and thoroughly accustomed to the management of animals. The use of packs, after the manner of our western mountaineers, was little known here, and I therefore gave it my personal supervision. In the course of six days everything was in readiness. One more man was needed to assist in taking care of the animals and packs, and for this purpose Macomber obtained for me the services of a native who was said to be more familiar with Mauna Loa than any person now living. He proved to be an excellent man, and of great assistance in many emergencies.

My packs contained a tent, blankets for the entire party, cooking utensils, six weeks' provisions, and a dry-plate photographic apparatus. It also contained two sheet-iron vessels for carrying water, each having a capacity of about eight gallons.

The first journey was to Kilauea, the distance from Waiohinu being by road about forty-five miles. The entire route was full of interest and instruction, and it may be well to note some of the more striking details by the way.

It is to be remembered that the long and gentle slopes of Mauna Loa are merely the surface of a mass of lavas which have been piled over each other in the form of lava streams poured out at intervals throughout an epoch of vast but unknown duration. These great lava floods burst out seemingly in a most capricious manner here, there, and everywhere. They break out far more frequently at or near the summit than upon the lower flanks of the mountain, and so vast is the amount of lava outpoured at each eruption that the streams often reach literally from the summit to the sea, spreading out from a quarter of a mile to two or three miles in width. Upon such a broad surface as that of Mauna Loa it must necessarily happen that some portions may lie for many centuries unscathed by fire, and during this period of immunity the lavas decay, soil is formed upon them, and accumulates to the depth of many feet. The district of Kau, therefore, exhibits many wide expanses of comparatively recent lava streams with intervals which have escaped the devastating flow of lava for so long a period that they are now deeply buried in soil. Upon such an interval is located the large Naalehu sugar plantation, which extends for a space of about four miles from the village of Waiohinu to the village of Honuapo. The land here is from 600 to 800 feet above the sea, and descends by a very gentle slope towards the ocean, suddenly ending in lofty cliffs which plunge at once into deep water. A little distance inland the country rises suddenly by steep slopes to a height of about 1,800 feet, forming bold and well-rounded hills. These hills are, for the most part, composed of soil, holding, however, intercalated beds of lava which crop out here and there. The principal mass of these hills is soil. If we ascend their summits we find ourselves, at a height of about 1,800 feet above the sea, upon a broad terrace which has a gentle, almost imperceptible slope upwards towards the mountain. I shall have frequent occasion to allude to this peculiar mode of formation, for it has an important bearing upon certain conclusions to which I have been led concerning the progressive upheaval of the southern part of Hawaii. The same formation occurs around the entire southern part of the island, though it is often more or less masked by recent lavas. The general idea which I wish to convey is that as the slopes of Mauna Loa approach the sea they disclose two and sometimes several terraces, one from 1,800 to 2,500 feet high, the other from 500 to 1,000 feet high. These terraces are composed of alternations of heavy beds of soil and lava sheets, the masses of soil largely predominating. We shall see more of these terraces as we proceed on our journey.

A good road leads from Waiohinu to Honuapo, the distance being about five miles. Through about two-thirds of that distance it runs along the terrace, at a height of 600 to 800 feet above the sea, and at length winds down the hillside to the beach at Honuapo. As we descend the hill we have upon our right a vertical cliff formed by the waves driven against it by the trade-wind. Directly in front of us, to the northeastward, are the long slopes of Kilauea descending, by a very gentle declivity, to the ocean, and projected against the horizon, 30 or 40 miles away. Here was pointed out to me a scarcely perceptible break or jog in the far distant profile of that slope, and I was told that it was the brink of the western wall which looks down into the great amphitheater of Kilauea. When night comes the observer may from this point behold the heavens aglow with the fires of that wonderful place.

As we reach the bottom of the hill we find ourselves upon a low platform a few feet above sea-level. The terraces now recede inland two or three miles, and present themselves under forms which are quite novel and striking. The view of them is somewhat distant, but is more comprehensive for that very reason. The eye of the traveler will be caught instantly by some large hills of singular form situated about four miles to the northeastward of Honuapo. Any one who has seen the mesas and buttes of the Plateau Country of the Rocky Mountain region will be instantly struck by the resemblance which these hills offer. In reality they are veritable buttes, having the flat summits, the steep slopes, with the angular and rectilinear ground plans, so often seen in the western regions of our own country. These masses are separated from each other by wide valleys having every appearance of valleys of erosion. Down these valleys vast streams of lava have descended, spreading out in wide fields over the low plain which lies between the terraces and the sea. Near the ocean they are black, gloomy, and horrible in appearance, exhibiting a roughness which, when beheld for the first time, impresses the beholder with a feeling which is half amazement, half amusement. As the eye follows these streams upward toward the source whence they came, it perceives that vegetation gradually appears, becoming more and more abundant the further upward it is followed, until the blackness of the stream is finally lost to sight in the mantle of living green. It is always interesting to note that those portions of the recent lava beds which are situated in a dry climate remain for very long periods without a trace of vegetation, while the portions which are kept moist by rain are clothed with herbage and even trees after the lapse of a very few years. He may observe, also, how the stream becomes very narrow where the slope is steep, but as soon it strikes a very gentle declivity it spreads out to tenfold or even twenty-fold width, and attains increased thickness.

The plain to the east of Honuapo and between the recent lava flows is quite smooth. It is floored with more ancient lava, which is only half covered by a film of soil a few inches in thickness. Here it is necessary to describe the two strongly-contrasted forms presented by lava streams on these islands.

The first form is called by the Hawaiians pa-hó-e-hó-e. Its general character can be appreciated far better by a drawing or photograph than by verbal description. Imagine an army of giants bringing to a common dumping-ground enormous caldrons of pitch and turning them upside down, allowing the pitch to run out, some running together, some being poured over preceding discharges, and the whole being finally left to solidify. The individuality of each vessel full of pitch might be half preserved, half obliterated. The surface of the entire accumulation would be embossed and rolling, by reason of the multiplicity of the component masses, but each mass by itself would be slightly wrinkled, yet, on the whole, smooth, involving no further impediment to progress over it than the labor of going up and down the smooth-surfaced hummocks. We always considered journeying over pahoehoe as good traveling by comparison.

The second form of the lavas is called by the natives a-a, and its contrast with pahoehoe is about the greatest imaginable. It consists mainly of clinkers sometimes detached, sometimes partially agglutinated together with a bristling array of sharp, jagged, angular fragments of a compact character projecting up through them. The aspect of one of these aa streams is repellent to the last degree, and may without exaggeration be termed horrible. For one who has never seen it, it is difficult to conceive such superlative roughness.

Why the same lava stream should in some portions of its extent take the form of pahoehoe and in others take the form of aa seems at first mysterious; but the explanation is not difficult, and it will appear very clearly when we come to examine the varying conditions which attend the flow of lavas and the circumstances under which they finally cool and solidify. When these lavas are discharged they come up out of the ground in enormous volumes, are intensely heated, and are very liquid. The vents being situated high up on the mountain where the slopes are considerable, they at first descend with immense velocity, running fifteen or twenty miles per hour, and so copious is the supply that they often reach a distance of eight or ten miles from the orifice before they feel very sensibly the effect of cooling. Wherever the slope is steep they run with great velocity. Where the declivity is less the velocity greatly diminishes and the flood spreads out over wider areas. As they become cooler they become more viscous. The cooling takes place upon the surface of the mass while the interior still remains hot and preserves a viscous liquidity. The superficial crust of cooled lava undergoes rupture at numberless points, and little rivulets of lava are shot out under pressure. Preserving their liquidity for a short time, they spread out very thin and are quickly cooled, forming pahoehoe. Scarcely is one of these little offshoots of lava cooled when it is overflowed by another and similar one, and this process is repeated over and over again. In a word, pahoehoe is formed by small offsoots of very hot and highly liquid lava from the main stream driven out laterally or in advance of it in a succession of small belches. These spread out very thin, cool quickly, and attain a stable form before they are covered by succeeding belches of the same sort.

The fields of aa are formed by the flowing of large masses of lava while in a condition approaching that of solidification. Ordinarily they are very thick and cover a very large area, representing therefore an enormous mass comparable to that of a glacier. The movement is in some respects glacier-like, but with this difference; instead of having a sensibly constant temperature throughout, it is hot within and more or less viscous and nearly cooled on the surface. A mass so large but still plastic undergoes, for a considerable time, a slow movement amounting to perhaps a few hundred yards or even a few hundred feet in a single day, gradually becoming slower until at last it ceases. During this slow glacier-like motion crushing strains of great intensity are set up throughout the entire mass, and its behavior conforms strictly to that of viscous bodies. The superficial portions in part yield plastically to the strains, in part yield by crushing, splintering, and fissuring. The result is a chaos of angular fragments. The aa streams are always thicker than the pahoehoe and rise from 50 to 80 feet above them, being bounded by a margin of low cliff and talus.

As already remarked, the same lava stream may exhibit pahoehoe or aa, according to the circumstances attending the flow, and the final form which the stream takes is quite independent of the chemical constitution of the lava. Excellent illustrations of both kinds may be found right here at Honuapo, and, in truth, there are no better contrasts of the two methods of solidification upon the island. About a mile from the village, as we proceed eastward, is an immense stream of quite recent aa a mile and a half in width and extending inland to one of the valleys between the terrace buttes before alluded to. A very good trail has been macadamized across the field, which would otherwise be literally impassable. As we traverse it we are deeply impressed with its rough, horrible aspect. Although this flow has an appearance of extreme recency the natives have no tradition concerning the period of its outbreak. No doubt it is more than a century old, and very likely more than two centuries.

The trail after leaving this stream descends again upon older pahoehoe partially covered with thin soil, and this continues for a little more than a mile, when it ascends a second field of aa broader and more rugged if possible than before. It has also the same aspect of recency, though tradition is equally silent concerning its age. From the summit of this stream the eye again reverts northward to the terraces, now a little more than two miles distant. Their aspect and form are now more striking than before. It was here that their meaning first began to dawn upon me. Right at the base of the most striking one, as well as upon its slopes and summit, is situated what is known as the Hilea plantation. In the latter part of my journey I visited this locality, where I was most hospitably entertained by the manager of the plantation. I examined these buttes and terraces and found them to consist of masses of soil, with intercalary sheets of lava, identical in all essential respects with the terraces at Naalehu and Waiohinu. Perhaps this is as good a place as any in which to discuss them and their meaning. (See Pl. IV.)

The alluvial material which constitutes the greater part of these buttes is mostly fine and contains but little coarse fragmental matter. The deposition of such material as surface alluvium is governed by perfectly definite laws, which are now pretty well understood.* The fragmental material produced by the decay of rocks is gathered by the streams and carried downward. This *débris* is in all stages of comminution from the finest silt to coarse rubble and bowlders. The power of the stream to transport it depends upon the velocity of the running water, and this velocity in turn depends upon the declivity of the bed. This declivity varies along its length in an irregular manner, but on the whole, or taking averages, the declivity of any stream becomes greater the higher up we go towards its sources. Therefore its ability to transport *débris* is greater above and less below. Hence, too, the *débris* which it gathers and carries easily in its upper reaches is in part dropped or deposited lower down. Thus we find the explanation of the generally observed fact that streams tend to cut down their channels in their upper courses and often build up their beds by deposition lower down. We may be very sure therefore that when a stream deposits at any place considerable quantities of fine detritus its gradient at that place must be more feeble. Now the buttes at Hilea stand from two and a half to three miles from the sea, and the lower ones are about 1,200 feet high. Under

* We must, of course, be sure in the first place that the materials under discussion are really alluvial. Soil often forms *in situ* by the decomposition of lava beds, and instances of it are abundant on Maui and Oahu, some of which will be spoken of hereafter. Some of the soils so formed might be easily mistaken for alluvium, though careful examination of good exposures will generally decide the question. In the present instance there is abundant evidence of the alluvial origin of the deposit.

the conditions now existing it is obviously impossible that these large masses of fine *débris* could have been deposited as alluvium. The average gradient between them and the sea is far too great. Some other state of affairs quite different to that we now see must have existed when they were built up. Nor is the required condition hard to suggest. If we assume that the land has risen 1,200 to 2,000 feet the difficulty vanishes. We should then find the running water checked by its confluence with the sea at these levels and able to deposit sediment in abundance. In further support of this inference let us look at the valleys which separate the buttes. These are like any other valleys of erosion, and represent presumably just so much material removed from the intervals between buttes. The cause of the valleys becomes plain. The rise of the land would at once give greater declivity to the streams seawards, and the process of alluvial deposition would give place to one of excavation. The streams would cut down their beds into the soft earth, and the rains would wash the sides of the cuts, gradually widening them out into broad valleys. This is the process of butte formation, exemplified in thousands of clear cases in the Plateau country and in many other regions. The buttes are mere remnants of a large alluvial formation which was originally continuous.

The evidence of such an occurrence as the one supposed, if it be a veritable occurrence, connects itself with many other facts, and the force of the evidence of its verity is chiefly in its cumulative character, like that of the reality of a glacial period. The more we see of this country the more will the evidences accumulate that these buttes are silent witnesses of an extensive upheaval of this part of the island at an epoch not very remote.

The terrace of which these buttes are remnants is seen with more or less distinctness all around the southern part of Hawaii, except in those places where it has been completely buried by many lava streams. Sometimes it is so much ravaged by erosion that the mind fails in the effort to restore it imaginatively. But at the right places we are sure to find it, even though it is sometimes battered and mutilated. The existence of such a terrace, indeed of two or three of them, raises at once the conviction of some common cause for them, and the cause suggested meets the problem exactly.

It is difficult to estimate with precision the amount of elevation attested by these terraces, but there are evidences still legible of several of them—one of them 1,200 to 1,400 feet high, another about 2,800, and perhaps, though more doubtful, a third at 3,400 feet. Behind the foremost buttes of Hilea may be seen the remnants of an older and loftier terrace in a much more advanced stage of decay. This upper terrace is so far wasted in this vicinity that the geologist might hesitate to draw any conclusion from the remnants. But further to the northeast it presents itself in a more significant form, and the evidence becomes more coherent.

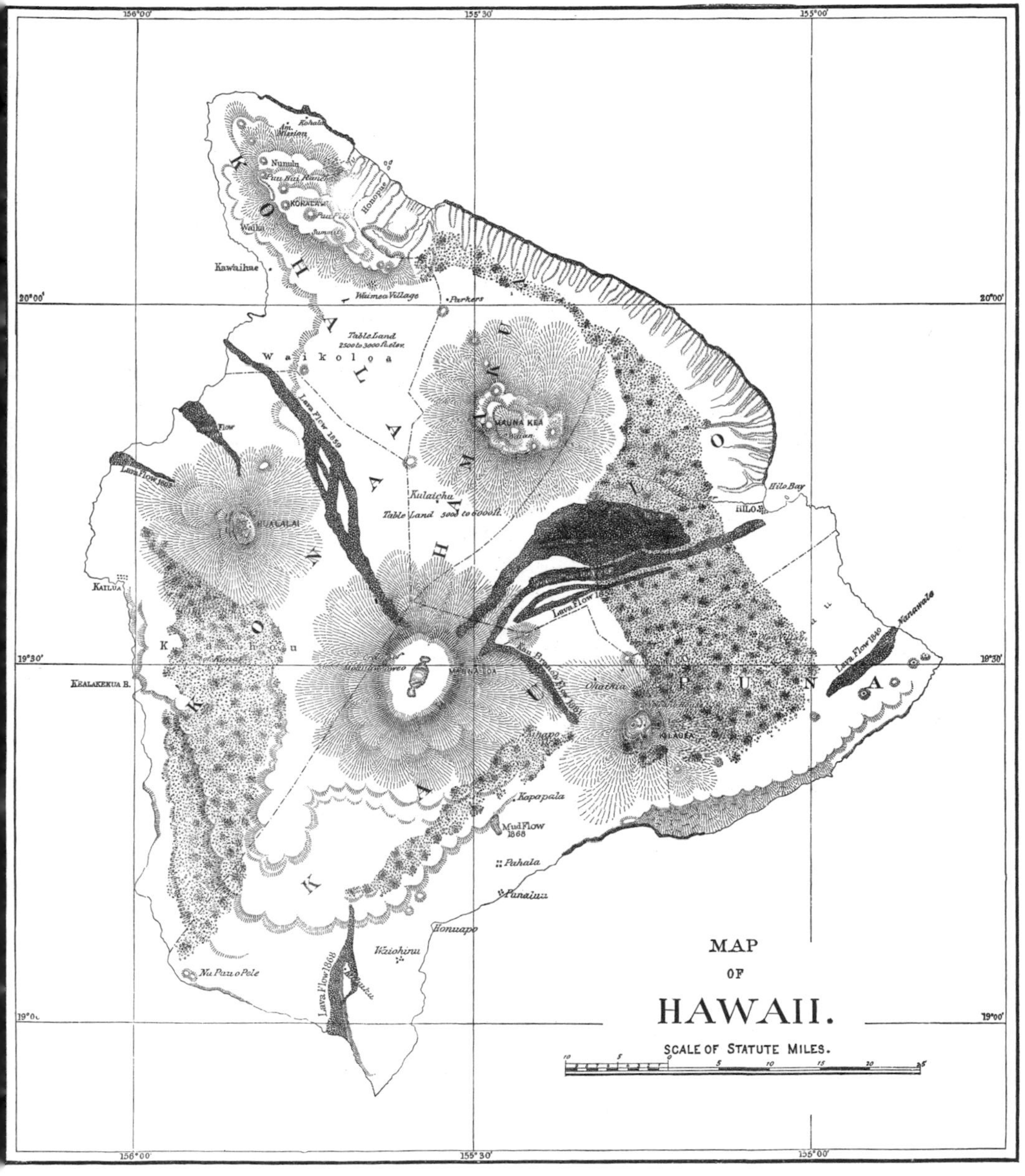

MAP

OF

HAWAII.

SCALE OF STATUTE MILES.

BUTTES AND TERRACES

EA—A FIELD OF AA.

PAHOEHOE—MAUNA LOA.

Descending from the second field of aa, the road again reaches the beach at a little village called Punaluu. Just where the trail strikes bottom is a little pool containing tule-grass and water-cresses. The water is slightly brackish, and a stream is seen emerging from a tunnel in the rocks at about the level of mean tide. Since this is a common occurrence and has much significance, it may be used as a text for a short sermon upon the drainage of Mauna Loa. Upon all the broad surface of this mountain there is hardly a living stream, however small. Yet the amount of precipitation is very great; but the rain instantly sinks into the rocks, and only during the most copious showers and storms does it gather even temporarily into floods or rivulets. The waters escape to the sea by subterranean passages. As we shall see further on, the mountain mass is full of such passages, and some of them, no doubt, receive the drainage. The escape of the waters is much like what would happen if water were poured upon a pile of loose sand or gravel resting upon an impervious platform. It would sink into the interstices and escape around the base of the pile. It is so with Mauna Loa. At very many points around its base and just at the level of the sea may be noted these escaping streams. Many of these subterranean rivulets find access to the ocean below sea-level, and I have frequently been told of places along the coast of Hawaii where the natives take their canoes or swim out from shore and plunge down to the bottom to obtain fresh water. I have also been told that a species of sturgeon is caught near the southern cape of Hawaii in a hundred fathoms of water, where the natives assert that it is only brackish. They also assert that this fish is caught nowhere else.

At Punaluu the road leaves the sea-coast and extends inland, ascending by a long and gentle slope. It lies over pahoehoe partially covered by thin soil. At several points it crosses shallow channels scored in the rocks by streams which run only during the heaviest rains. It is well to note such facts, for they show us that although there are no perennial streams upon the mountain as yet, there are spasmodic ones, and whenever the volcano becomes extinct these may at length become permanent water channels, unless they are overflowed by future eruptions.

About five miles from Punaluu is the Pahala plantation. It is situated upon the lower portions of the alluvial terrace. Here the thinly covered lava platform gives place to great depths of fine, rich soil. Towards the mountain the alluvial land rises rapidly with a few intercalary masses of lava. The terrace is here much broken down and sloped off by erosion, but its lineaments are still perfectly distinguishable, and its character is not to be mistaken. Winding onward through the plantation for a distance of about three miles and steadily ascending, we at last reach the summit of the lower terrace, at a height of about 1,600 to 1,800 feet above the sea. Its surface is somewhat uneven, and the altitudes at different points may vary as much as 200 feet. Still further inland the higher terrace now presents itself more conspicuously than heretofore, and in a very striking manner. Its height is about 1,000 feet above the lower terrace. Its upper platform is sensibly smooth to the eye, while its face descends by a very steep slope. It is composed principally of alluvial soil containing beds of lava.

Here the road deflects to the right a little, and leads along the surface of the lower terrace and about a mile from the foot of the upper one. It soon crosses what is known to the inhabitants by the name of the great mud-flow, which took place in the year 1868, during the prevalence of a period of terrible earthquakes, accompanied by an immense wave from the sea rolling in upon the southern coast, and also by one of the great eruptions of Mauna Loa upon the southwestern side of the mountain, about thirty miles from here. The scenes which occurred during that momentous period were described to me by several people who witnessed them, and an account of them will be given in another chapter. The so-called mud-flow of Kapapala took place during the culminating part of this earthquake period, and was very probably started by one of the severest shocks. The principal alluvial terrace at this point is composed of unconsolidated clay, and it was saturated with water supplied by springs. The face of the terrace is very steep as well as very lofty, and it is easy to understand how the mass saturated with water might have been put in motion by a severe earthquake shock. Certain it is that a great land-slip occurred, and an enormous mass of clay containing some gravel and a few small bowlders detached itself from the great bank and flowed about two miles and a half, with great swiftness and in a stream about a third of a mile in width. So sudden and swift was the movement that the people of a native village directly in front of it who saw it coming were overwhelmed by it, and about thirty of them were buried. Not an individual who was in its way had time to escape, and those who were spared were such as happened at the time to be in positions which the flow did not touch. So far as I can learn, no evidence of heat or volcanic action attended this catastrophe, although some accounts (which I do not credit) represent that it was accompanied by exploding steam. It appears, however, to have been a genuine land-slip detached by an earthquake, the materials composing it having been rendered sufficiently fluent by saturation with water. For several months the mud stream was so soft that it was impassable. At present it is overgrown by rank grass which affords excellent pasturage. Its thickness appears to be between forty and sixty feet.

A little beyond is situated what is known as the Kapapala ranch. It is one of those localities which seem to possess local importance and interest for which no sufficient reason can be assigned. Probably, however, its notability may be associated with the fact that it is the last watering-place on the road to Kilauea, and is also a halting-place for travelers who wish to ascend Mauna Loa. The traditional hospitality of the natives to whom it originally belonged, and of the white proprietors who have succeeded them, is, no doubt, a still more intelligible reason. It also has an importance somewhat analogous to that of some of our western hamlets where several lines of railroad intersect, for at this point several divergent roads lead to as many different parts of the island. Here, too, we begin to pass into a tract of lavas which appear to be associated in part at least, if not chiefly, with Kilauea as distinguished from Mauna Loa.

It is customary to speak of Kilauea as a mere appendage of Mauna Loa and situated upon its flanks. As my familiarity with the relations of the two increased, it produced a growing impression of distinctness in the two volcanoes. As we approach Kilauea from the Kau side this impression of distinctness will, I think, become stronger and stronger.

From Kapapala the road winds on, slowly and steadily ascending over ancient fields of pahoehoe thinly covered with soil. The lavas here have buried the lower terrace, and now and then an ancient flow from Mauna Loa is seen descending from the upper terrace, merging with the lavas below. Half a mile from the ranch the trail passes a deep cavity in the ground, and the rocks ring hollow beneath the feet. It is a common occurrence, and one which we might have noted before, because we have passed many such already on our route. It is an old lava pipe. A lava stream which has been flowing for several days gradually forms an outer covering by the superficial cooling of the lava, making a regular tunnel. Probably no great eruption takes place without the formation of several such tunnels, perhaps many of them. They are often of great extent and even as much as three or four miles in length. Here and there the roof of the tunnel falls in. Sometimes a single slab drops in, forming a skylight for the cavern below. More frequently the tunnel preserves its arch. There are literally thousands of these tunnels throughout the mass of Mauna Loa. Their transverse dimensions are highly variable, sometimes expanding into a great chamber 60 or 80 feet in height and of corresponding width, again contracting to an aperture of a few square yards. So numerous are these caverns that it seems as if they must form some appreciable part of the entire volume of the mountain.

The lavas we are traversing in the vicinity of Kapapala no doubt originated from Mauna Loa. But a little to the right and seaward is a barren wilderness of black lava which certainly originated in chief part at least from the purlieus of Kilauea; for at a distance of about four miles from Kapapala the trail descends upon this plain, reaching a spot from which the general surface again ascends in all directions except to the southwest. Mauna Loa is upon our left and the long declivity of Kilauea is upon our right, and we ride along a line where the two conical surfaces intersect. Undoubtedly the lavas from the two sources are blended together and alternate with each other. Upon our right also, at a distance of about three miles, may be seen four or five cinder cones standing upon a line which, if prolonged, would pass into the great basin of Kilauea. These cones are conspicuous rather for their rarity than for anything else. The geologist who has rambled much among the scenes of recent volcanism will greatly miss these almost invariable accompaniments of activity, and wonder at their paucity. The line of cones seen upon our right is situated upon a fissure, the prolongation of which carries it directly into the pit of Kilauea. There are several fissures traversing this great lava-plain, all radiating from Kilauea, and from some of them volumes of steam are still issuing. This is a noteworthy fact, pointing to the individuality of Kilauea as a distinct volcanic center. We shall find that similar fissures radiate from the summit of Mauna Loa.

And now for a time the trail winds pleasantly along upon a grassy bottom of soil with the lava beds upon our right and an alluvial bank upon our left. This alluvial bank, no doubt, is a degraded exposure of one of the alluvial terraces hitherto noted, which just here has escaped burial by recent lavas. At length the trail leaves the alluvial bottom and runs into the broad fields of naked pahoehoe. Around us and reaching southward and eastward into the dim distance is a barren desolate waste of rolling and smoothly rounded hummocks of solid rock. Once, no doubt, they were black as coal, but now they are faintly brownish or ruddy from weathering. The path winds tortuously among them, now making a detour to escape some shattered pile or ragged crack, now tumbling over a wrinkled, contorted knoll of solidified lava.

The first impression produced by the sight of one of these vast fields of naked lava is very memorable. It has something akin to the first prospect of the sea or the great plains, or an arctic ice-field. It conveys a sense of grandeur, solemnity, desolation, but above all, monotony. Very impressive, too, is the sense of magnitude and power which it produces. Rarely are such widespreading lava wastes to be found elsewhere in the world. Probably those of Iceland equal them, and those of the Snake River country incomparably surpass them, but I know of no others of equal magnitude. The journey over them is monotonous and wearisome, for the rolling surface for the most part shows no more diversity than the prairies of Iowa or a newly planted corn-field. And yet there is one diversion. After a few miles of pahoehoe we find ourselves in front of an ugly, ominous barrier, which scowls and bristles across the path as if to forbid a nearer approach to the Inferno beyond. It is the edge of a great field of aa, stretching for many a weary mile across the broad expanse of rolling pahoehoe. A good trail has been macadamized across it with a course as straight as an arrow. It has the ordinary aspect with which we have already become familiar in the great fields of aa between Honuapo and Punaluu. It seems, however, to be somewhat older than those, or rather it has been more discolored by weathering, which, however, is not a safe guide to inferences concerning age, since the amount of weathering depends altogether upon climate. This field of aa came from a prehistoric though doubtless recent eruption, from a vent situated upon the northeastern flank of Mauna Loa, at an altitude of about 8,000 feet, and about seventeen or eighteen miles from the summit. It rivaled in magnitude the great historic eruptions, and it eventually reached the sea on the Kau coast about five miles east of Punaluu. The entire length of the stream was about twenty-three miles. The upper portion of it is composed largely of pahoehoe. But where it strikes this phlegrean field on which we are traveling the very feeble slope checked its velocity, enabling it to spread out and to flow sluggishly. In accordance with the mechanism already described, it here takes the form of aa. But further on to the southward, where the slope again becomes much greater, the stream narrows, and for the most part takes the form of pahoehoe. Here at its narrowest part the field of aa is a little over a mile in width, and its thickness is probably between 60 and 80 feet. There is no tradition indicating the time of its eruption.

Descending the eastern wall of this field of aa we are once more upon a vast expanse of pahoehoe. The journey now becomes toilsome. The hummocks of lava are large and high and the animals lurch and strain as they scramble over them. But as the surface in detail is otherwise smooth the hardship is limited to severe work, the foothold being perfectly secure. For about four miles the trail keeps near the barrier of aa, winding among the hummocks of pahoehoe. At length it deflects away from the aa and points as straight as possible for Kilauea. The ascent is very gradual and it is only by consulting the barometer that we become conscious how rapidly we are gaining in altitude. After about twelve miles of floundering among these hummocks we find ourselves at the foot of a rather steep hill which is broken away on the right by an abrupt cliff. As we ascend it, the trail, rising out of a little rain gully, leads us to a narrow platform. In front of us the surface of the earth instantly drops from the face of a vertical wall about 500 feet high, and Kilauea is before us.

CHAPTER III.

KILAUEA.

The Kau trail first strikes the edge of the Kilauea amphitheater upon its western side, and, following the western rim, circles around the northern end until it reaches the Volcano House. A few hundred yards beyond this point, where the volcano first breaks into view, we reach, by a sharp acclivity, the loftiest point of the encircling crest-line of the amphitheater. It is a memorable spot. Behind us rises the dome of Mauna Loa, and nowhere else upon the island is the superlative grandeur of this king of volcanoes displayed to such advantage. When the curtain of clouds is drawn aside, we behold also far to the northward the almost equally majestic mass of Mauna Kea. In front of us and right beneath our feet, over the crest of a nearly vertical wall, more than 700 feet below, is outspread the broad floor of the far-famed Kilauea. It is a pit about three and a half miles in length and two and a half miles in width, nearly elliptical in plan, and surrounded with cliffs for the most part inaccessible to human foot, and varying in altitude from a little more than 300 feet to a little more than 700 feet. The altitude of the point on which we stand is about 4,200 feet above the sea. The object upon which the attention is instantly fixed is a large chaotic pile of rocks, situated in the center of the amphitheater, rising to a height which by an eye estimate appears to be about 350 to 400 feet. From innumerable places in its mass volumes of steam are poured forth and borne away to the leeward by the trade wind. The color of the pile is intensely black, spotted and streaked here and there with red—not the red of fire, but of iron persalts, alternating with the magnetic black. Its general form is conical, holding a large crateriform depression within. But it is so shattered and broken that it has a craggy, ominous aspect, which may well be called hideous. Around it spreads out the slightly undulated floor of the amphitheater, as black as midnight. To the left of the steaming pile is an opening in the floor of the crater, within which we behold the ruddy gleams of boiling lava. From numerous points in the surrounding floor clouds of steam issue forth and melt away in the steady flow of the wind. The vapors issue most copiously from an area situated to the right of the central pile and in the southern portion of the amphitheater. Desolation and horror reign supreme. The engirdling walls everywhere hedge it in. But upon their summits and upon the receding platform beyond are all the wealth and luxuriance of tropical vegetation heightening the contrast with the desolation below.

Yet we can pause here but a few moments. The journey has been long and wearisome and we must seek rest and shelter, in order to survey the scene with deliberation. A ride of two miles further brings us to the Volcano House, which is a very comfortable hostelry, affording much needed shelter, for we are almost sure to reach it in the midst of a driving rain. The entire distance from the Kau coast to the verge of the amphitheater lies through a country which is almost arid. And yet as soon as we reach the summit point of the surrounding wall, we pass, in the space of half a mile, from a very dry region to a very moist one. These rapid transitions from wet to dry and *vice versâ* are common on the island of Hawaii. And nowhere is this transition more abrupt than at Kilauea. The transverse axis of the pit may be used as a sharply dividing line between two strongly contrasted climates. North of it the rainfall is excessive. South of it the rainfall is very deficient.

I have hitherto carefully avoided applying the term "crater" to Kilauea. It has so little in common with the orifices from which lavas and other volcanic products emanate that the word is little more than a misnomer here. All the accessories of Kilauea differ immensely from those associated with normal craters, and it seems necessary to apply to it some designation expressing its distinct characters. Counterparts of Kilauea are not common. I can think of but one in the Mediterranean volcanoes which appears to be at all homologous to it, and that is the Val del Bove; though my acquaintance with European volcanoes is insufficient to enable me to say whether or not this is the only one in Europe. On these islands the exact counterpart of Kilauea is the great pit on the summit of Mauna Loa named Mokuaweoweo. The vastly larger depression on the summit of Haleakala, though differing much in form, is in my opinion strictly homologous. Numerous small crateriform depressions are found in many parts of Hawaii, which also seem to me to be homologous with Kilauea, some of which are only a few hundred feet in diameter, and none of them exhibit any signs of recent activity. Considered with reference to their origin the evidence is conclusive that they were formed by the dropping of a block of the mountain crust which once covered a reservoir of lava, this reservoir being tapped and drained by eruptions occurring at much lower levels. A great deal of misconception and confusion of ideas have arisen from the practice of characterizing these depressions by the term craters. For example, few people speak of Haleakala without calling it the greatest crater in the world. My understanding is that it is not a crater at all, and that we have therefore nothing to compare it with except such formations as I have just mentioned. It seems necessary therefore to adopt some term which will apply to the very limited class of amphitheaters of which Kilauea may be considered as the type. Perhaps the term *caldera* may be as unobjectionable as any, though I am far from regarding it as quite satisfactory, and am fully prepared to find it severely criticised. It is the best one I can offer, and hereafter I shall employ it in speaking of these formations wherever they occur in these islands.

The morning after reaching the Volcano House, I descended to the floor of Kilauea. The wall of the caldera at its northern end has settled into a series of steps by the sinking of successive portions which have faulted off from the main platform of the country. This dropping of successive spalls of great size, many hundreds of yards in length and from 20 to 200 yards in width, is conspicuous around many portions of the parapet. Each of these great spalls forms a shelf or ledge, backed by a steep and sometimes vertical cliff. Wherever the escarpment is sloped a good foot trail has been dug, allowing of a steep but safe and easy descent. By such a trail the floor of Kilauea is reached without difficulty. As soon as we reach the bottom we find ourselves upon brand new pahoehoe of the most typical kind. We travel over rolling, smooth-surfaced bosses of rock without difficulty for a distance of about a mile and three-quarters, when we reach a rapidly ascending slope, which rises a little more than a hundred feet. Gaining the summit, we find ourselves upon the brink of a pool of burning lava. This pool is about 480 feet long and a little over 300 feet in width. Its shape is reniform. It is surrounded by vertical walls 15 to 20 feet in height. When we first reach it the probabilities are that the surface of the lake is coated over with a black, solidified crust, showing a rim of fire all around its edge. At numerous points at the edge of the crust jets of fire are seen spouting upwards, throwing up a spray of glowing lava drops and emitting a dull, simmering sound. The heat for the time being is not intense. Now and then a fountain breaks out in the middle of the lake and boils feebly for a few minutes. It then becomes quiet, but only to renew the operation at some other point. Gradually the spurting and fretting at the edges augment. A belch of lava is thrown up here and there to the height of 5 or 6 feet, and falls back upon the crust. Presently, and near the edge, a cake of the crust cracks off, and one edge of it bending downwards descends beneath the lava, and the whole cake disappears, disclosing a naked surface of liquid fire. Again it coats over and turns black. This operation is repeated edgewise at some other part of the lake. Suddenly a network of cracks shoots through the entire crust. Piece after piece of it turns its edge downward and sinks with a grand commotion, leaving the whole pool a single expanse of liquid lava. The lake surges feebly for a while, but soon comes to rest. The heat is now insupportable, and for a time it is necessary to withdraw from the immediate brink. Gradually the surface darkens with the formation of a new crust, which grows blacker and blacker until the last ray of incandescence disappears. This alternation of the freezing of the surface of the lake and the break-up and sinking of the crust goes on in a continuous round, with an approach to a regular period of about two hours. The interval between the break-ups varies, so far as observed, from forty minutes to two hours and a quarter. Probably the average interval is somewhat less than two hours. The explanation of the phenomenon is not difficult. It is now believed that lavas at temperatures a little below the point of solidification have a specific gravity slightly less than that of liquid lava; but when they have cooled considerably below the point of congelation their specific gravity is greater in the solid than in the liquid state. Hence, when the crust first forms it is light enough to float, and is very thin, but it gradually thickens, and the upper part grows denser. At length a stage is reached at which the mean density of the crust is greater than that of the lavas beneath, and the position of the crust then becomes unstable. This instability is first shown at the edges of the lake, where the mechanical support of the crust is interrupted by the escaping gases and by the feeble boiling of the liquid mass within. As the density of the crust increases the strain set up by the yielding at the edges at length becomes sufficient to propagate itself through the whole mass and break it up into fragments, which at once sink.*

When the lava is freshly exposed in the lake it has exactly the appearance of melted cast iron, its color shading from red through orange into yellow. It is easy to see that the temperature of the freshly exposed surface is by no means so great as many have been led to suppose. Basic lavas are very fusible, and those before us have the appearance of being decidedly viscous and sluggish in their movements. It is very probable that the basalts which come up out of the earth in

* Since the above attempt to explain the periodic break-ups was put in writing I have felt distrustful of it. As these pages are undergoing revision I therefore take advantage of the opportunity to amend it. It seems to me that the conclusion that lavas expand in solidifying is open to question, and that the experiments from which it is derived may have been vitiated by an omission to take account of the following facts. It is certain that the melted silicates readily occlude notable quantities of water, and when they solidify they exclude the water just as water itself excludes air in freezing. The excluded gas or vapor, however, is mechanically entangled in the solidifying mass in the form of bubbles or vesicles. As these vesicles are often minute, they may have been neglected and no account taken of them. But as their number is vast they may have seriously diminished the apparent density, in the case of lava or cast iron, though it is not to be supposed that this would affect the main fact (expansion in solidifying) in the case of water.

Now, the silicates, when passing from the melted to the solid condition, pass through a very considerable range of temperature within which they are viscous. If the lava be kept for a long time, say an hour or two, well within this range of temperature the steam vesicles would have an opportunity to disentangle themselves from the mass and escape just as they do in a glass furnace. The amendment, therefore, which I would offer is as follows: The first inch or two of crust which forms is cooled quickly and becomes stiff and black in a few minutes. Its *absolute* density is presumably greater than that of the liquid below; but, being full of vesicles and spongy, it is light enough to float. Subsequent additions to its thickness are made to its under surface. Each successive film so added has a longer and longer time in which to disengage its gaseous contents. Therefore they successively become more and more compact, and their successive specific gravities approach more and more nearly the absolute density of the substance. Hence, as the thickness of the crust increases, its specific gravity increases. When the mean specific gravity becomes greater than that of the liquid (as it surely does) the position is unstable, and rupture once started is quickly propagated through the entire crust, which goes to pieces and sinks.

great eruptions reach the surface at a far higher temperature than those seen in this lake, being, in fact, most probably at a white instead of a red or yellow heat.

The phenomenon of Pele's hair is often spoken of in the school books, and receives its name from this locality. It has generally been explained as the result of the action of the wind upon minute threads of lava drawn out by the spurting up of boiling lava. Nothing of the sort was seen here, and yet Pele's hair was seen forming in great abundance. Whenever the surface of the liquid lava was exposed during the break-up the air above the lake was filled with these cobwebs, but there was no spurting or apparent boiling on the exposed surface. The explanation of the phenomenon which I would offer is as follows: Liquid lava coming up from the depths always contains more or less water, which it gives off slowly and by degrees, in much the same way as champagne gives off carbonic acid when the bottle is uncorked. Water-vapor is held in the liquid lava by some affinity similar to chemical affinity, and though it escapes ultimately, yet it is surrendered by the lava with reluctance so long as the lava remains liquid. But when the lava solidifies the water is expelled much more energetically, and the water-vapor separates in the form of minute vesicles. Since the congelation of all siliceous compounds is a passage from a liquid condition through an intermediate stage of viscosity to final solidity, the walls of these vesicles are capable of being drawn out as in the case of glass. The commotion set up by the descending crust produces eddies and numberless currents in the surface of the lava. These vesicles are drawn out on the surface of the current with exceeding tenuity, producing myriads of minute filaments, and the air, agitated by the intense heat at the surface of the pool, readily lifts them and wafts them away. It forms almost wholly at the time of the break-up. The air is then full of it. Yet I saw no spouting or sputtering, but only the eddying of the lava like water in the wake of a ship. The country to the leeward of Kilauea shows an abundance of Pele's hair, and it may be gathered by the barrel-full. A bunch of it is much like finely shredded asbestos.

The lava pool before us is called the New Lake. It was formed in May, 1881. It opened suddenly in the floor of the caldera, and was at first of much smaller dimensions than at present. It has been gradually enlarged by the cleaving off and engulfment of successive slices of its encircling wall. In truth the lava lakes do not, as a rule, maintain any constant position. They have been seen to open in various parts of the floor, and after some months or years of activity, similar to that already described, they freeze up permanently, and are entirely obliterated. Five or six lakes were known in 1853. Four years ago (1878) a large open lake existed near the north end of the pit, but its location is no longer descernible. In truth, the floor of the caldera is liable to open and become a lava pool at almost any point. The changes have been very great within historic times, and some recital will be be made presently of its condition in the first part of the present century.

There is a second and larger lake presenting a somewhat more dramatic appearance. It is situated less than half a mile from the New Lake, within the large chaotic pile of rocks or cone which first attracted our attention when we reached the brink of the caldera. It has occupied its present situation many years, so long in fact that the imperfect records kept of its changes hardly permit us to form an idea of the exact period of its origin. Its name is Halemaumau. This name was applied to the great central lake of Kilauea, as it was first seen by Ellis in 1823, and Halemaumau may be fairly considered as representing what now remains of that great feature. As we pass from the New Lake to Halemaumau, we have abundant evidence that we are treading upon the thin crust of a slumbering volcano. Numerous cracks on either hand emit steam and sulphurous vapors. The rocks are corroded and chemically changed by the action of acid gases, and are warm to the touch upon the surface. A stick plunged into one of these cracks is quickly charred. Signs of instability, such as shattering and heaving movements, are seen all around, and the thoughts of a catastrophe are ever before the mind. At length we reach the encircling cone which surrounds the older lake. At the north end the barrier is broken down into a mass of rubble and sharp angular fragments, over which progress is somewhat difficult. Ascending a steep slope of lava fragments, we soon reach the summit of it, and the lake is before us. We cannot, however, approach it as we could the New Lake. A vertical cliff, at the foot of which is a series of yawning cracks and fissures, sending out intensely hot steam and the most acrid vapors, forms an insuperable barrier. Still we may command from an elevation of about a hundred feet a very good view of the greater part of the pool. Its aspect is somewhat different from that of the New Lake. There is more activity, and its surface is covered with boiling fountains of liquid lava, but none of them spout to any great altitude. In the presence of this ebullition a thick crust like that of the New Lake cannot form. The surface is too unquiet. Still, the greater part of the lake is covered with a thin black crust which floats in detached sheets which sink from time to time. The periodical changes and alternations between congelation and the sinking of the crust are not so well marked. From time to time periods of comparative quiet supervene, followed by periods of general activity throughout the pool. The area of Halemaumau also is larger than that of the New Lake, being nearly 1,000 feet in length, with a width of nearly 600 feet.

The cone which surrounds this lake is a very striking construction, or rather destruction. It is not composed of fragmental material like an ordinary crater, but of masses of lava which have been apparently pushed up. The elevatory movement has been accompanied with much shattering and contortion, and the rocks have been thrown into such attitudes that it seems as if a breath would knock the whole thing down. Still it is a definite structure, having some features in common with the summit cone of Vesuvius. It consists, in fact, of cones within cones. That it has really been hoisted is testified by those who have occupied the Volcano House since 1875, and the greater part of the upheaval has taken place within the last three years.

The amount of steam and gaseous exhalations is very much greater in Halemaumau than in the New Lake. The ebullition in the former is constant, while in the latter it is very feeble and almost insignificant. Fountains of liquid lava rising to the height of 5 to 10 feet, as nearly as could be judged by an eye estimate, are seen at all times in Halemaumau. But they are generally confined to a few localities at any given time. They change about frequently from place to place, breaking out suddenly in one spot and gradually dying away in another. The amount of condensed steam floating away in the form of white vapor is not so very great, when we consider the very large surface from which it emanates. Most of this visible steam, however, does not come apparently from the surface of the lava itself but from the fissures and numberless vent holes in the wall of the surrounding crater. It is probable, however, that the intense heat of radiation from the pool itself prevents the condensation of the steam until it has diffused itself throughout a considerable body of the atmosphere. Over the entire surface of the burning lake is spread a pall of translucent vapor, through which the remoter wall of the crater is still visible, though somewhat clouded. What proportion of these vaporous exhalations consists of sulphur gases it is impossible to estimate, though I have no doubt that it is considerable. Some of it is probably in the form of sublimed sulphur, which collects in small quantities upon the leeward side of the crater. Some of it may perhaps be anhydrous sulphuric acid. But as no precipitate or deposit of this acid has been detected, I have no better ground for this conjecture than the fact that some of the white fumes did not appear to have the odor either of sublimed sulphur or of sulphurous acid, being far more acrid. The presence of hydrochloric acid would have been readily detected by its peculiar odor, which is quite unmistakable. Still there are some indications that hydrochloric acid is among the exhalations of this volcano. In many places the lava is bleached by the abstraction of its iron protoxide. In many small spots are seen brilliant red, orange, and saffron colors. The bleaching would be most readily accomplished by hydrochloric acid, and similar red and orange spots are known to be produced by the conversion of iron chloride so formed into peroxide.

In general, there is not disclosed to the eye an amount of condensed vapors which seems to be adequate to account for the amount of ebullition taking place over the surface of the lake, and it is quite probable that much of the vaporous products are carried off by the wind in the form of invisible vapor. To these gaseous emanations we may look for an explanation of the persistent liquidity of the lava within the lake.

They rise, no doubt, from very great depths in the form of bubbles at a temperature very much higher than that of the lava at the surface of the lake and replace all the heat which is lost at the surface by radiation. In their ascent it is probable that they produce also convection currents consisting of ascending eddies of hotter lava and descending eddies of cooler lava.

It seems necessary, however, to pursue our inquiries a little further in this direction. At great depths below the surface we may presume the vapors to have a much higher temperature than at the surface. But they are also under enormous pressure, and as they ascend in the lava column the diminishing pressure ought to be accompanied by an elastic expansion of the vapors, and this in turn would lower the temperature by reason of heat becoming latent. Thus it may happen that unless these vapors are most intensely heated below and have temperatures very greatly above that of the surface the loss of temperature by this expansion would render them incapable of imparting more heat to the superficial portions of the lava column. No estimate of this possible loss of temperature, however, is practicable. But the convection currents would not be liable to this criticism.

The earlier visitors to Kilauea whose accounts of it are now accessible speak of a phenomenon which did not exist at the time of my visit. I refer here to what have been termed "blowing cones" within the lake. Ellis, in his account of Kilauea in 1823, describes them as "conical inverted funnels" rising to heights varying from 20 to 40, or even 50 feet above the surface of the lake, with openings at the top from which jets of vapor and sometimes spouts of lava were thrown out. As many as fifty were seen at one time within the great lava lake then existing, and most of them were simultaneously active. The same phenomenon was described in 1825 by parties from the H. B. M. frigate Blonde. They were also seen by Wilkes in 1841, and have frequently been seen within the last ten or fifteen years by many other visitors. They appear to have been composed of solidified but very hot lava. None of them were permanent, but after a short period of activity they were either melted down or shifted their positions. Ultimately, no doubt, they were remelted. That they shifted their positions is fully attested by many observers. Most probably they were masses of solidified lava floating like bergs in the lake. During my visit two masses of solid lava formed within the New Lake. These had no orifices at their summits and showed no action at all suggestive of blowing cones. They appeared to be simply masses of solidified lava formed out of the pool itself. In the course of several days these islands, as we called them, certainly shifted their positions by a very considerable amount, one of them moving across two-thirds of the shorter diameter of the lake. We may perhaps account for their buoyancy by the supposition that lava at a temperature a little below that of congelation is specifically lighter than lava a little above that temperature. There is difficulty, however, in

VIEW OF KILAUEA

VOLCANO HOUSE.

THE NEW LAVA LAKE—KILAUEA.

HALEMAUMAU—THE GREAT LAVA LAKE.

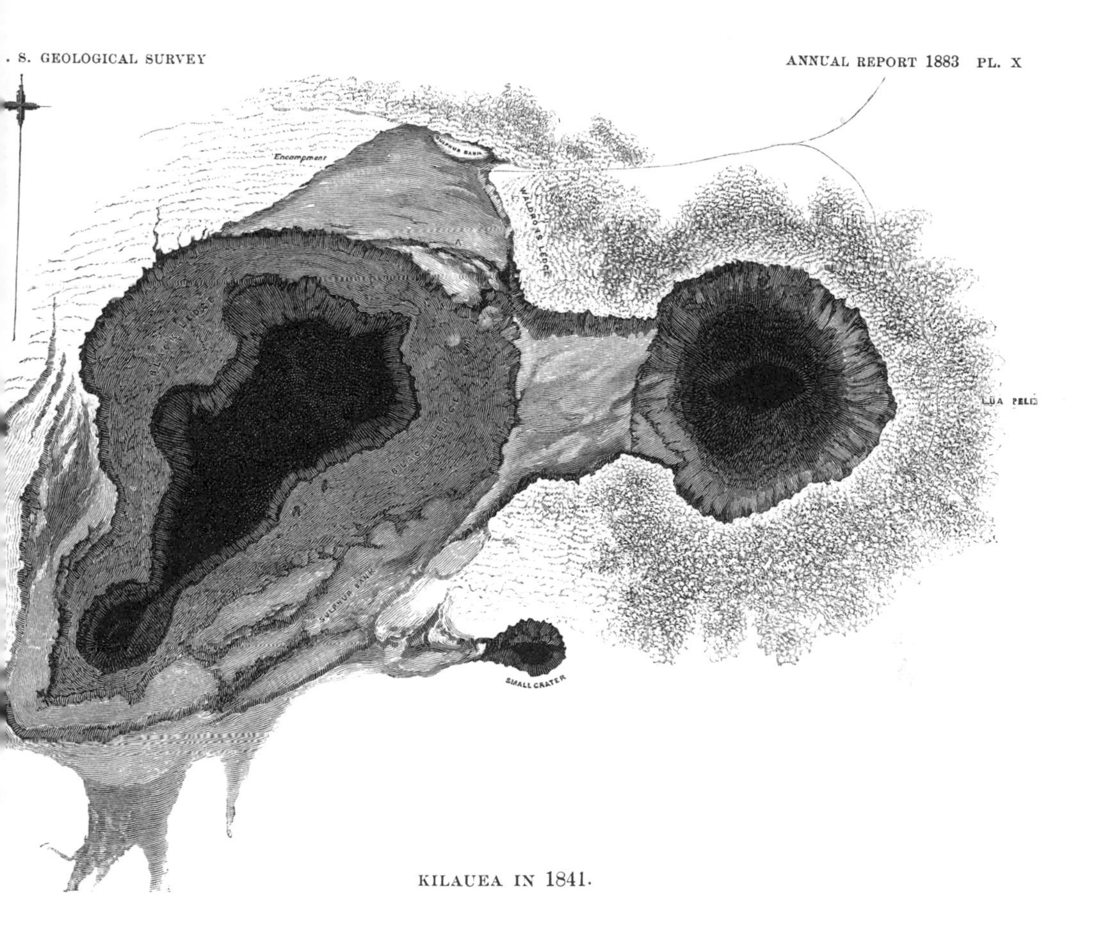

KILAUEA IN 1841.

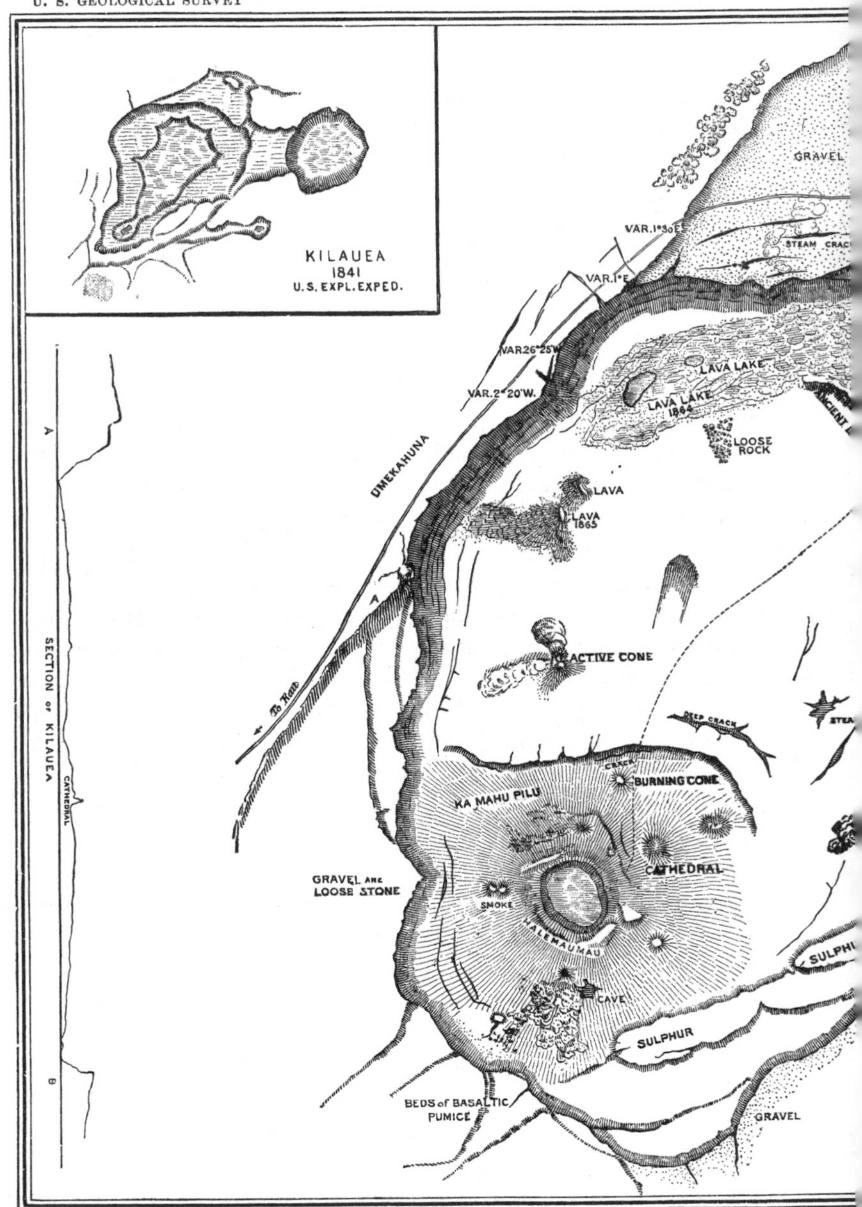

KILAUEA
1841
U.S. EXPL. EXPED.
SECTION of KILAUEA
A
B
CATHEDRAL
GRAVEL
STEAM CRAC
VAR.1°30'E.
VAR.1°E.
VAR.26°25'W.
VAR.2°20'W.
LAVA LAKE
LAVA LAKE
1864
LOOSE
ROCK
UMEKAHUNA
LAVA
LAVA
1865
A
To Kau
ACTIVE CONE
DEEP CRACK
CRACK
BURNING CONE
KA MAHU PILU
CATHEDRAL
GRAVEL AND
LOOSE STONE
SMOKE
HALEMAUMAU
CAVE
SULPHUR
SULPHU
BEDS of BASALTIC
PUMICE
GRAVEL

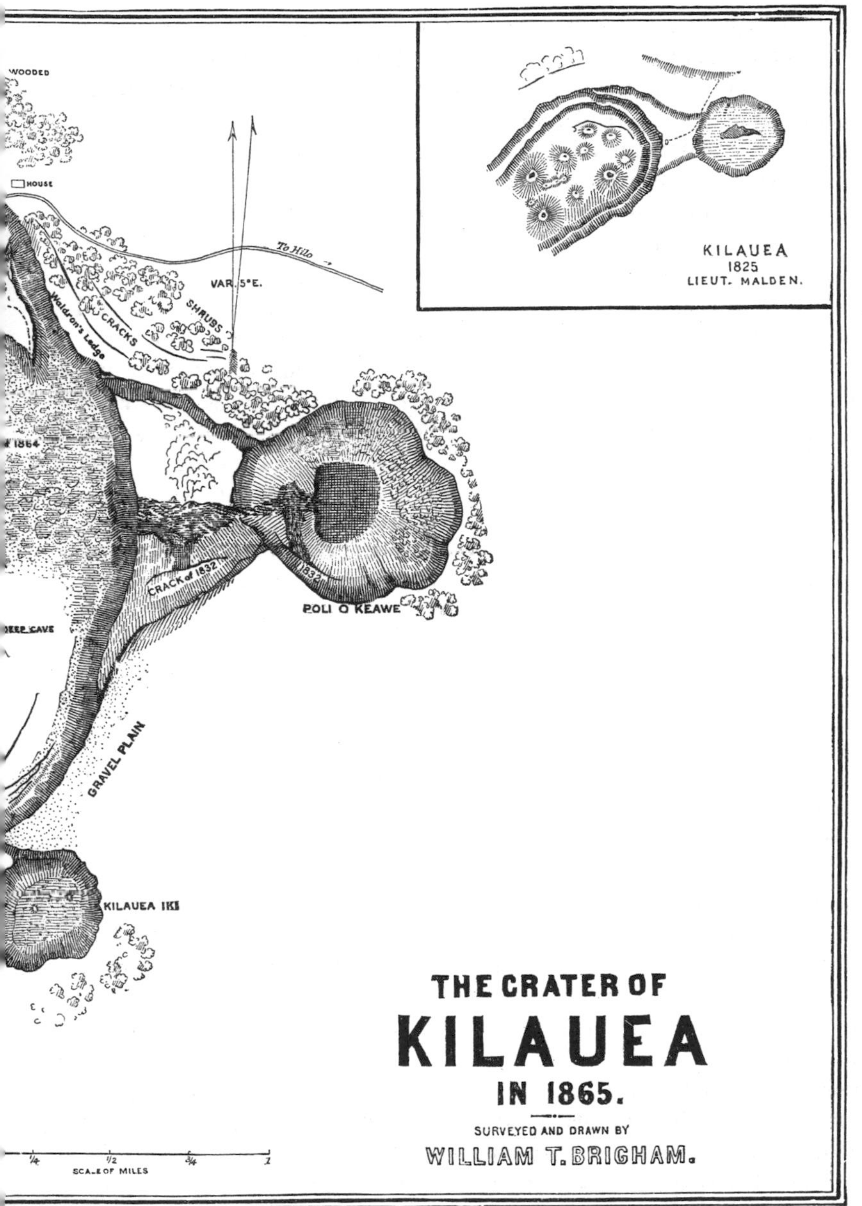
WOODED
HOUSE
To Hilo
VAR. 5°E.
SHRUBS
CRACKS
Waldron's Ledge
1864
CRACK of 1832
1832
POLI O KEAWE
DEEP CAVE
GRAVEL PLAIN
KILAUEA IKI
1/4
1/2
3/4
1
SCALE OF MILES
KILAUEA
1825
LIEUT. MALDEN.
THE CRATER OF
KILAUEA
IN 1865.
SURVEYED AND DRAWN BY
WILLIAM T. BRIGHAM.

understanding how a considerable mass of lava could so congeal within the pool at all. The principal mass of lava surrounding it is still in a liquid condition. I have been unable to find any satisfactory explanation of this problem.

Leaving Halemaumau and passing around the rocky cone which incloses it, we may enter the southern half of Kilauea. Half a mile to the southwest of the cone the aspect of the floor of the caldera becomes very repellent. Great quantities of steam and bluish vapor rise from innumerable rifts and cracks, and blending into a dense cloud, float away to the leeward. Here existed a few years ago a large lava lake, which is now entirely frozen over, though the clouds of steam still indicate plainly the thinness of the covering. This pool of lava was known by the name of the Old South Lake. Although it has ceased to exist as a distinct lake, it still emits at intervals of a few months or even a few weeks considerable quantities of lava, which overflow portions of the adjoining floor. It requires some courage to venture upon an area so dangerous, though in reality the risk of accident is not very great, if ordinary caution is exercised. Very little knowledge, however, is to be gained by such a journey, for it is impossible to visit the places we would most like to inspect on account of the great quantities of pungent gases emanating from the numberless fissures. The surface consists wholly of pahoehoe, which has an unusually spongy and vesicular character, and which crumbles beneath the feet. Innumerable blisters occur, which break beneath the tread and let the foot down into holes, from which it must be instantly withdrawn to prevent the shoe from being burned to a crisp. The black color of the lava has been discharged in many places by the reaction of acid vapors, here changed to a snowy white, there to an orange, red, or saffron color.

The Old South Lake sealed up about three years ago, but the eruptions from it have been frequent since that time, though most of them have been quite small.

It is interesting to recur to the accounts of Kilauea given by those who visited it in the early part of the century. Prior to the discovery of the islands by Captain Cook (1776), we have no accounts excepting the most fanciful myths. There is, however, a tradition which was learned by the earliest missionaries concerning an eruption of Kilauea, which is supposed to have taken place in the year 1789, and to which a certain amount of credence is given by the most intelligent among the earlier writers. This account relates that Keoua, King of Hilo, being at war with the King of Kau, marched his army in three divisions past Kilauea; that while the columns were in movement a violent eruption took place, during which great volumes of smoke were shot high in the air, carrying with them great quantities of rocks and hot stones, and that one division of the army was destroyed to the last man by the inhalation of the sulphurous vapors. As this event took place during the reign of Kamehameha I, and must have been fully within the recollection of many people living at the time of the advent of the missionaries, and as it is very circumstantial in its account, there may be justification for the belief that an eruption of an unusual and violent character took place at that time. It is certain, however, that no subsequent eruption has been attended with the like degree of violence, or with action resembling in character that which the tradition describes. All primitive peoples are addicted to the grossest exaggeration, and are quite incapable of describing any natural phenomenon with accuracy. The most that we are at liberty to infer from this account is that an eruption of great violence may have taken place in that year attended with occurrences which have not since repeated themselves.

In the year 1823 the Rev. William Ellis, an English missionary, made an extended tour of the island of Hawaii, and reached Kilauea from Kau by way of Kapapala, the same route which has been described in the preceding chapter. The description which he gives leads to the belief that Kilauea presented at that time an aspect differing greatly from the present one. He says:

"We found ourselves on the edge of a steep precipice with a vast plain before us 15 or 16 miles in circumference and sunk from 200 to 400 feet below its original level. The surface of this plain was uneven, and strewed over with large stones and volcanic rocks, and in the center of it was the great crater, at the distance of a mile and a half from the precipice on which we were standing. Descending into the crater and walking some distance over the sunken plain, which in several places sounded hollow under our feet, we at length came to the edge of a great crater, where a spectacle sublime and even appalling presented itself before us. Immediately before us yawned an immense gulf in the form of a crescent about two miles in length from northeast to southwest, nearly a mile in width, and apparently 800 feet deep. The bottom was covered with lava, and the southeast, northeast, and northern parts of it were one vast flood of burning matter in a state of terrific ebullition, rolling to and fro its fiery surge and flaming billows. Fifty-one conical islands of varied form and size, containing as many craters, rose either around the edge or from the surface of the burning lake, twenty-two constantly emitting columns of gray smoke or pyramids of brilliant flame, and several of these at the same time vomiting from their ignited mouths streams of lava which rolled in blazing torrents down their black indented sides into the boiling mass below. The sides of the gulf before us, although composed of different strata of ancient lava, were perpendicular for about 400 feet and rose over a wide horizontal ledge of solid black lava of irregular breadth, but extending completely around. Beneath this ledge the sides slope gradually towards the burning lake, which was, as nearly as we could judge, 300 or 400 feet lower. It was evident that the large crater had been recently filled with liquid lava up to this black ledge, and had by some subterranean canal emptied itself upon the sea or upon low land on the shore; and in all probability this evacuation had caused the inundation of the Kapapala coast, which took place, as we afterwards learned, about three weeks prior to our visit. The gray, and in some places apparently calcined sides of the great crater before us; the fissures which intersected the surface of the plain on which we were standing; the long banks of sulphur on the opposite side of the abyss; the vigorous action of the numerous small craters on its borders; the dense columns of vapor and smoke that rose at the north and south end of the plain, together with the range of steep rocks by which it was surrounded, rising probably in some places 300 or 400 feet in perpendicular height, presented an immense volcanic panorama, the effect of which was greatly augmented by the constant roaring of the vast furnaces below.

* * * * * * *

"Between 9 and 10 in the evening the dark clouds and lava fog that since the setting of the sun had hung over the volcano gradually cleared away, and the fires of Kilauea, darting their fierce light athwart the midnight gloom, unfolded a sight terrible and sublime beyond all we had yet seen.

"The agitated mass of liquid lava, like a flood of melted metal, raged with tumultuous whirl. The lively flame that danced over its undulating surface, tinged with sulphurous blue or glowing with mineral red, cast a broad glare of dazzling light on the indented sides of the insulated craters, whose roaring mouths, amidst rising flames and eddying streams of fire, shot up at frequent intervals, with very loud detonations, spherical masses of fusing lava or bright ignited stones.

* * * * * * *

"As eight of the natives with us belonged to the adjoining district, we asked them to tell us what they knew of the history of this volcano and what their opinions were respecting it. From their account and that of others with whom we had conversed we learned that it had been burning from time immemorial, or, to use their own words, from chaos until now, and had overflowed some part of the country during the reign of every king that had governed in Hawaii; that in earlier ages it used to boil up, overflow its banks, and inundate the adjacent country; but that for many kings' reigns past it had kept below the level of the surrounding plain, continually extending its surface and increasing its depth, and occasionally throwing up with violent explosion huge rocks or red-hot stones. These eruptions, they said, were always accompanied by dreadful earthquakes, loud claps of thunder, with vivid and quick succeeding lightning. No great explosion, they added, had taken place since the days of Keoua; but many places near the sea had since been overflowed, on which occasions they supposed Pele went by the road under ground from her house in the crater to the shore."

I quote this last paragraph, not because I attach much weight to primitive traditions, but because the statements it contains seem so intrinsically probable. There is no evidence that the lavas have, within any recent period, overflowed the outer rim of the caldera. On the contrary, the fields of pahoehoe, which now form the crests of the surrounding walls, are quite ancient and are considerably decomposed by many centuries of weathering. Yet there is reason to believe that at some ancient epoch such outflows actually occurred, because the streams of lava in some instances, perhaps in many instances, radiate away from the rim of the caldera. That the great pit has progressively enlarged its dimensions through a considerable period of time is also most probable, for there is strong evidence that this enlargement is still going on from time to time by the sinking of large spalls or slices which break off from the walls of the surrounding precipice. The black ledge referred to in Ellis's description was at that time, no doubt, one of the most striking features of Kilauea. It is described a few years later in some detail by Mr. Stuart, who accompanied Lord Byron, commanding H. B. M. frigate Blonde. In Lord Byron's narrative Kilauea is figured in drawings made by Lieutenant Malden, R. N., and his sketch shows the great interior pit and the black ledge as very conspicuous features.

The Wilkes Exploring Expedition visited these islands in 1840–'41, and Lieutenant Wilkes has given a very excellent account of the condition of Kilauea at that time. He says:

"When the edge is reached, the extent of the cavity becomes apparent, and its depth became sensible by comparison with the figures of some of our party who had already descended. The vastness thus made sensible transfixes the mind with astonishment, and every instant the impression of grandeur and magnitude increases. To give an idea of its capacity, the city of New York might be placed within it, and when at its bottom would hardly be noticed, for it is three and a half miles long, two and a half wide, and over a thousand feet deep. A black ledge surrounds it at the depth of 660 feet, and thence to the bottom is 384 feet. The bottom looks in the day-time like a heap of smouldering ruins. The descent to the ledge appears to the sight a short and easy task, but it takes an hour to accomplish.

"All usual ideas of volcanic craters are dissipated upon seeing this. There is no elevated cone, no igneous matter or rocks ejected beyond the rim. The banks appear as if built of massive blocks which are in places clothed with ferns nourished by the issuing vâpors. What is wonderful in the day becomes ten times more so at night. The immense pool of cherry-red liquid lava in a state of violent ebullition illuminates the whole expanse and flows in all directions like water, while the illuminated cloud hangs over it like a vast canopy.

"We sat on its northern bank for a long time in silence until one of the party proposed we should endeavor to reach the bank nearest to and over the lake; and having placed ourselves under the direction of Mr. Drayton, we followed him along the edge of the western bank; but although he had been over the ground the day before, he now lost his way and we found ourselves still on the upper bank, after walking two or three miles. We then resolved to return to the first place that appeared suitable for making a descent, and at last one was found, which, however, proved steep and rugged. In the darkness we got many a fall and received numerous bruises; but we were too near the point of our destination to turn back without fully satisfying our curiosity. We finally reached the second ledge and soon came to the edge of it. We were then directly over the pool or lake of fire, at the distance of about five hundred feet above it, and the light was so strong that it enabled me to read the smallest print. This pool is 1,500 feet long by 1,000 feet wide, and of an oval figure.

"I was struck with the absence of any noise except a low murmuring like that which is heard from the boiling of a thick liquid. The ebullition was (as is the case where the heat is applied to one side of a vessel) most violent near the northern side. The vapor and steam that were constantly escaping were so rarified as not to impede the view, and only became visible in the bright cloud above us which seemed to sink and rise alternately. We occasionally perceived stones or masses of red hot matter ejected to the height of 70 feet and falling back into the lake again. The apparent flow to its southern part is only because the ebullition on the north side causes it to be higher, and the waves it produces consequently pass over to the opposite side.

"The black ledge is of various widths, from 600 to 2,000 feet. It extends all around the cavity, but it is seldom possible to pass around that portion of it near the burning lake, not only on account of the stifling fumes, but of the intense heat. In returning from the neighborhood of the lake to the point where we began the ascent we were one hour and ten minutes of what we considered hard walking; and in another hour we reached the top of the bank. This will probably give the best idea of its extent and the distance to be passed over in the ascent from the black ledge, which was found to be 660 feet below the rim. Messrs. Waldron and Drayton finally reached the floor of the crater. This was afterwards found to be 384 feet below the black ledge, making the whole depth 987 feet below the northern rim. Like the black ledge it was not found to have the level and even surface it had appeared from above to possess; hillocks and ridges from 20 to 30 feet high reached across it and were in some places so perpendicular as to render it difficult to pass over it. The distance they traversed below was deceptive, and they had no means of ascertaining it, but by the time it took to walk it, which was upwards of two hours from the north extreme of the bottom to the margin of the large lake. It is extremely difficult to reach this lake on account of its overflowing at short intervals, which does not allow the fluid mass time to cool. The nearest approach that any one of the party made to it at this time was about 1,500 or 2,000 feet. They were then near enough to burn their shoes and light their stakes in the lava which had overflowed during the preceding night.

"The smaller lake was well viewed from a slight eminence. This lake was slightly in action; the globules (if large masses of red fluid lava several tons in weight can be so called) were seen heaving up at regular intervals six or eight feet in height; and smaller ones were thrown up to a much greater elevation. At the distance of 50 feet no gases were to be seen nor was any steam evident; yet a thin smoke-like vapor rose from the whole fluid surface; no puffs of smoke were perceived at any time.

"At first it seemed quite possible to pass over the congealed surface of the lake to within reach of the fluid, though the spot on which they stood was so hot as to require their sticks to be laid down to stand on. This idea was not long indulged in, for in a short time the fluid mass began to enlarge; presently a portion would crack and exhibit a bright red glare; then in a few moments the red lava stream would issue through and a portion would speedily split off and suddenly disappear in the liquid mass. This kind of action went on until the lake had extended itself to its outer bank and had approached to within 15 feet of their quarters, when the guide said it was high time to make a retreat. The usual course is for the lake to boil over, discharge a certain mass, and then sink again within its limits. It is rarely seen to run over for more than a day at a time."

Similar accounts of the condition of Kilauea between the years 1823 and 1841 have been given by other parties who visited it during that period. The most striking feature at that time must have been the great inner cavity of the caldera and the surrounding black ledge. This interior depression has now wholly disappeared. It has been filled up completely. And not only that, but the portions which it once occupied are built up so far that they now form the highest part of the floor of the main caldera. Over what was once the most active part of this great lake has been built up the choatic pile of crateriform rocks which now encircle the pool of Halemaumau. Both the north and the south lakes have disappeared, having been frozen over completely, and in their stead the New Lake, situated half a mile northeast of Halemaumau has made its appearance.

It appears that in 1841 the level of the liquid lava at the bottom of the great interior pit was a little more than a thousand feet below the highest point of the outer wall of the caldera. At the present time the level of the liquid lava in the New Lake is about 580 feet below the highest point of the rim, and its level in Halemaumau I judge to be about the same, but it was impossible to obtain access to it in order to verify this inference. At the present time, then, the liquid lava columns stand about 435 feet higher than they did forty years ago. No record has ever been kept of the progressive action by which these changes have been brought about. Nothing remains to show the successive steps in the accretion of lavas which gradually filled up the interior pit. The only guides we have are the fragmentary accounts of numberless visitors describing the condition of Kilauea from time to time. These are all so incoherent and so grossly wanting in precision that it is impossible to frame a connected account of the process. There are, however, a few general features of the process which appear, and these may be briefly summarized. All accounts go to show that the height of the liquid column oscillates in an irregular manner, and while most of these oscillations are small, usually not exceeding 10 to 15 feet, yet in exceptional cases they are very much greater. Whenever the liquid column rises there is a tendency to overflow the margin of the pool which surrounds it, and this frequently happens. The quantities of lava thus outflowing and spreading out over a considerable area vary extremely, being sufficient sometimes to cover no more than a few acres to the thickness of a very few feet, while on rare occasions a square mile or two may be overflowed with a considerable body. The duration of these overflows is also extremely variable. Sometimes it is a single belch or surge lasting but a few minutes. It is quite common for the lava to run in this way for a whole day, and in larger overflows it may run for two or three weeks without interruption. Sooner or later the liquid column sinks and the overflow ceases. The eruptions are not by any means confined to the lakes, but break out at unexpected places. One of the most favored spots for this action is the former locus of the Old South Lake, which for several years has been completely frozen over. The cooling lava invariably takes the form of pahoehoe.

In reading the earlier accounts of Kilauea and in comparing them with the condition prevailing at the time of my visit, I was at first impressed with the idea that there had been on the whole a decrease in the amount of volcanic energy within the last forty years or more. But a more careful and critical study of these accounts has tended to efface that impression. It is evident that the writers were profoundly impressed with the sublime spectacle, and their deepest emotions were stirred. It is natural under such circumstances that their writings should portray the intensity of their feelings. It is not to be suspected for a moment that they intended to exaggerate, but under the spur of intense enthusiasm they aimed rather to express what they felt than to give a rigorous and exact description in cool, formal language. The reader of these accounts also is apt to be somewhat at fault, for he is liable to be intent upon the dramatic aspect of the scene and to share the enthusiasm of the writers and to forget or be indifferent to the strictest exactitude. Scrutinizing these earlier accounts more closely it will be seen that there is little reason to suppose that the amount of surface of liquid lava exposed to the atmosphere was any greater then than now. The amount of energy displayed would depend entirely upon the quantity of vaporous products given off. And it is by no means certain that there has been any diminution in this respect. In truth, this form of energy is never constant. It increases and diminishes from week to week and month to month, and it is a common saying that Kilauea is never twice alike. At the time of my visit it was probably more quiet than usual. And yet, immediately after I left it, there broke out from a point near the New Lake one of the largest eruptions which has ever been known to take place within the caldera. The lava flowed steadily for nearly a month, and completely overflowed more than half of the floor of the caldera, profoundly changing the aspect of its details. The boiling and surging, the spouting upward of the lava in fountains was feeble and unfrequent during my visit; but it had occurred a short time before with great power, and there is no reason to doubt that it may occur still more forcibly hereafter. The phenomenon of the blowing cones, however, has not been witnessed for some years.

CHAPTER IV.

PURLIEUS OF KILAUEA.

Having examined the salient points of interest in the great caldera it will be instructive to take a broader and more general view of the volcanic pile of which Kilauea is the focus. It has been habitual on the part of almost all writers to speak of Kilauea as situated upon the flanks of Mauna Loa and forming merely an appendage to that mountain. Many considerations have led me to regard it as a distinct volcano, having no more connection with Mauna Loa than any other volcanic center of the Hawaiian group. The horizontal distance from Kilauea to the summit of Mauna Loa is about 19 miles. The distances from the summit of Mauna Loa to the summits of Mauna Kea and of Hualalai are respectively 22 and 20 miles. So far as the length of the interval is concerned it is quite sufficient for as great a degree of independence as that prevailing between any two adjacent volcanoes. The difference in altitude between the lava lakes of Kilauea and the central pool of Mokuaweoweo is about 9,300 feet. The idea of a liquid connection or continuity through subterranean passages between these lava lakes seems to be so thoroughly opposed to all hydrostatic laws as to be incredible upon the very face of it. It seems impossible that the two vents can derive their lavas from a common reservoir.

If we take our stand upon the western brink and highest point of the wall inclosing Kilauea, we shall observe that the profile of the country descends towards Mauna Loa. A slight but still decided depression exists towards the latter mountain. Across this depression a horizontal line will strike the nearest portion of Mauna Loa at a distance of about 4½ miles, and the intervening depression along this line amounts to about 340 feet. This indicates to us the fact that Kilauea is situated upon a totally distinct mountain pile. The distinction, however, has been in some measure masked and rendered inconspicuous from a variety of causes. In the first place, all the profiles or mountain slopes in this vicinity are exceedingly flat and weak. This is true both of Kilauea and Mauna Loa. Again, the space intervening between the two mountains is a region where the lavas from the two sources have in former periods overflowed each other and are now intercalated. But the gigantic floods from Mauna Loa have been emitted probably more frequently and in much larger volume, and, no doubt, constitute much the greater part of the lava masses occurring in this interval. Thus the two mountains have, so to speak, grown into each other. And if we might be permitted to look forward to an indefinite growth of the colossal pile of Mauna Loa, we might conceive of it as ultimately overgrowing and burying Kilauea completely. It is difficult here to form a purely mental conception of the enormous scalè upon which the mass of Mauna Loa has been constructed, or to imagine the immense spread and volume of its far-reaching flanks. Yet Kilauea has much the same character, being quite as flat in all its profiles, if not more so, and in proportion to its altitude, spreading out quite as broadly. The most generalized view which can be taken of Kilauea is that it is an exceedingly flat cone intersecting or adjacent to the much larger cone of Mauna Loa. As we become gradually acquainted with the topographical details of the country lying within 5 or 6 miles of Kilauea we at length become confident of the fact that it is a distinct cone or dome.

The caldera of Kilauea, however, is not situated at the apex of this independent mountain pile, but lies about four or five miles to the westward of the apex, and by just so much the nearer to Mauna Loa. It is apparent that it has not always been the center or focus of the volcanic activity of the mass. No doubt, at some former epoch this focus was situated at the apex, and it has been transferred to its present situation at an epoch which is presumably recent. Evidence of this may be found in riding around the pit and examining the lava beds which form the surface of the adjoining country. These radiate, for the most part, from the apex of the main cone, as if they flowed originally from that direction. Still other evidences may be found, of which the most striking are the occurrence of those singular abnormal, abortive cinder cones which stand over the cracks or fissures radiating from the summit of the pile. These radiating fissures are very characteristic both of Kilauea and Mauna Loa, and diverge from volcanic centers. In much more recent times a great fissure has been formed, starting from the southern end of Kilauea and reaching a distance of about 16 miles to the south-southwest. It is still open, and emits steam at many points, and was in that condition when seen by Ellis in 1823. But the older fissures, with here and there a cone formed above them, radiate from an apex four or five miles to the eastward.

The summit platform in the immediate vicinity of the caldera discloses numerous points of interest; and I propose to describe them as they present themselves to the observer who makes the circuit of Kilauea, taking the Volcano House as a starting point.

The first feature which will catch the attention of the geologist is the manner in which the platform in the vicinity of the pit is riven by faults. Large fragments of the wall cleaving off from the main platform have sunken to various depths; and this feature manifests itself around the entire circumference. The most complicated instance of this fracturing and sinking of detached blocks is presented along the trail descending from the Volcano House to the floor of the caldera. The courses of the faults are, with some notable exceptions, parallel to the rim of the surrounding walls. The magnitudes of the sunken blocks are always considerable, and in some cases their upper surfaces have areas which form a considerable fraction of a square mile. Those in the immediate vicin-

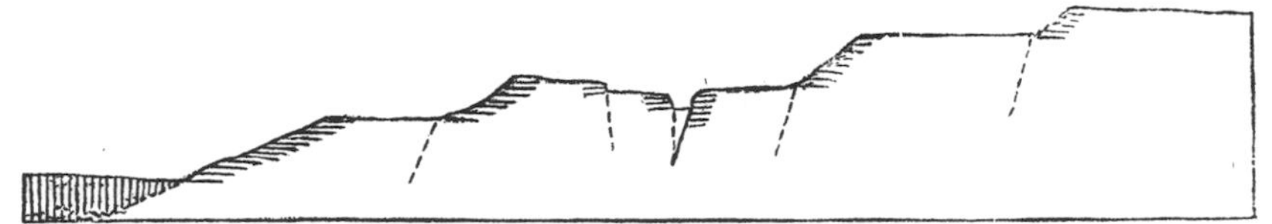

FIG. 3.—Faults in the northern wall of the caldera at Kilauea.

ity of the upper wall are, in some cases, nearly a mile in length, and vary in width from 20 or 30 yards to more than 300 yards. The accompanying section will exhibit in the most concise form the apparent relations of these faulted fragments in front of the Volcano House. (Fig. 3.)

Immediately west of the hotel a road descends about 100 feet upon a sunken platform of very large proportions. It is well figured on the map drawn by the Wilkes Exploring Expedition. Towards the south this platform ends upon the brink of Kilauea, and upon all other sides it is inclosed by ledges evidently originating in a circuitous fault, with displacements varying from 30 to 150 feet. Numerous fissures traverse this platform, having courses parallel to the brink of the caldera. They open in gaps from two to six feet in width and are of unknown depths. From nearly all of them hot steam issues, and it is evident that the heat is intense at depths of only a very few feet below the surface. From most of them nothing emanates but the vapor of water. But in that portion of the platform nearest to the Volcano House large quantities of sulphur vapor exhale which condense in the forms of arborescent and acicular crystals of sulphur which are very beautiful. At other places fumes of sulphurous acid are given off in abundance. It is not a little remarkable that these three classes of volcanic products should, as a general rule, be quite distinct. The larger portion of the emanations consists of steam which shows no trace of acid. Over the steam cracks ferns and bushes grow with such rank luxuriance that the crack is often concealed, and the traveler must be on his guard while walking through the shrubbery lest he be precipitated into one of them. The evidence is abundant that the volcanic fires underlying this sunken platform have a very slight depth.

Proceeding around the northern end of the pit, the ground steadily ascends upon the northwestern and western sides. Here, too, are ob-

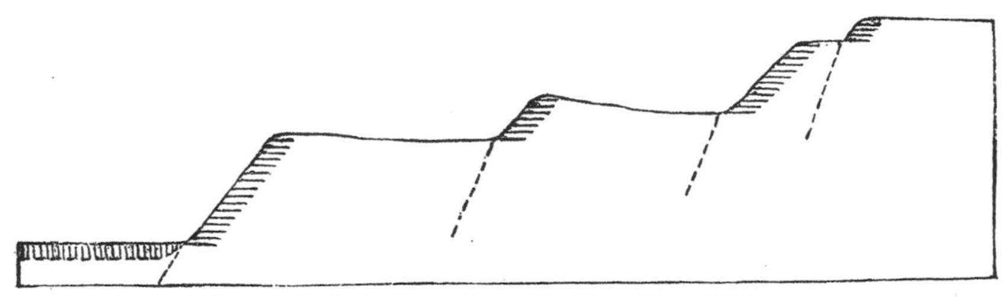

FIG. 4.—Faults in the western wall of the caldera at Kilauea.

servable many open cracks, which still maintain a general parallelism with the brink of the main precipice. At about the middle of the western side of Kilauea we reach the highest point of its surrounding wall, which has an elevation of about 730 feet above the floor in its vicinity. Here, too, the faulted and sunken blocks are shown very clearly and are of large proportions, being about a mile in length and from 100 to 350 yards in width. This is also the most commanding spot from which Kilauea may be viewed and the best idea obtained of its general features. Farther southward the wall declines in altitude, but still maintains its abruptness, being very nearly vertical. Passing around the southern end new features engage our attention. Notable among these are the fragmental products scattered over the ground. It should be remembered that the northern and eastern portions are upon the windward side, while the southern and western portions are upon the leeward. Whatever volcanic products may be cast up into the atmosphere and wafted away by the winds are to be seen upon the leeward, and never upon the windward border. Moreover, the leeward side is arid, while the opposite is very rainy and continuously clothed with vegetation. Hence these fragmental products are well preserved and well exposed. Among them are considerable quantities of coarse sand or fine lapilli occupying the swales between the bosses of pahoehoe or drifting about in little sand dunes. This material is seen in considerable abundance in coming from Kau. It has been swept by the infrequent rains into the low, broad washes and carried to the distance of eight or ten miles. It is supposed by many of the white residents that this black sand was thrown out in the traditional eruption of 1789, alluded to in the last chapter. Nor does the supposition seem at all improbable.

Another very common product here is a peculiar pumice. It is wonderfully light and spongy, and has a dark olive-green color. Many heaps of it are scattered about, consisting of small fragments seldom larger than a lemon, and usually smaller. It is the lightest pumice I have ever seen. Pele's hair is also very abundant, and may be gathered in wisps where it has been caught and held by some projecting fragment of rock.

At the south end of the pit we come suddenly upon a yawning fissure from 12 to 15 feet in width, extending indefinitely towards the south-southwest. This is known among the residents as the Sixteen-mile Crack, for it extends that distance away from the wall of the amphitheater in a nearly straight line. About a mile and a half from the brink two considerable cinder cones have been thrown up immediately over this fissure. At many points along its length clouds of steam escape and float away in the trade wind. Some search is necessary in order to find a place where this crack is narrow enough to be crossed. After passing it we still find the platform rifted in several places by small faults, some of which are parallel to the rim of Kilauea. Others radiate from it, so that the general arrangement is like that of a cobweb. None of them have any great width, but being concealed in places by drifted sands which form an insecure bridge, caution is necessary as we proceed.

Moving up the eastern side of the amphitheater, we at length come upon a large, deep pit, sunken in the platform at a distance of only a few hundred yards back of the crest of the main wall. Its diameter, judging by the eye, is about 1,500 or 1,600 feet, and the opening is approximately circular. The surrounding walls are nearly vertical most of the way down. It is known by the name of Kilauea-iki, or little Kilauea, and is one of a class of pits of which a considerable number are found upon the island of Hawaii.

Moving northward along the brink we come, at the distance of about a mile from Kilauea-iki, to a steep declivity or hillside trending away from the rim of the main amphitheater. About midway down this slope there is an old fissure from which in the year 1832 a small eruption of lava took place. The course of this fissure is nearly perpendicular to the main wall of Kilauea and opens along the hillside. Reaching the foot of the hill, a portion of the lava deflected to the west and poured down into the main amphitheater. Another portion of it deflected to the east, and poured down into a large pit situated about one-third of a mile from Kilauea and known by the name Poli-o-keawe. This pit is also nearly circular, with a diameter a little less than three-fourths of a mile, and a depth of about 750 feet, which is considerably below the floor of Kilauea. The neck or isthmus which separates it from Kilauea is also sunken below the level of the surrounding platform by about 200 feet. This sinkage has evidently taken place by faults radiating from the main amphitheater towards Poli-o-keawe.

The eruption of 1832 was a very small one, but is remarkable from the fact that the fissure from which it emanated opens at a level of more than 400 feet above the present lava lakes, and probably 700 or 800 feet above the lava lakes as they existed in 1832. So far as known, the lavas within Kilauea showed no sympathy with this eruption at the time it occurred.

Proceeding along the separating isthmus we encounter an exceedingly sharp, abrupt cliff upon the north end of the neck, which is somewhat difficult to mount, although its altitude above the neck is only about 240 feet. Reaching the upper platform we find it rifted with cracks, which give issue to steam. Their course is parallel to that of the wall. A further journey of a mile through the forest brings us back to the Volcano House.

Before proceeding to summarize our observations, it is desirable to note another series of important facts connected with Kilauea. In a line extending nearly eastward from the caldera is a continuous chain of pits and craters, reaching quite to the sea-coast, near the eastern extremity of the island. Along this line it is apparent that from time to time through the past centuries fissures have opened, disgorging lavas and building cinder cones. One of the most striking facts connected

POLI-O-KEAWE—NEAR KILAUEA.

BLOWING CONES.

(Reproduced from Ellis' Polynesian Research, 1823.)

with this action has been the formation of large circular pits similar to Kilauea-iki, and Poli-o-keawe. The natives residing in the Puna district have always averred that eruptions have taken place along this line with considerable frequency, and within the reign of every king of that district from time immemorial. The last one occurred in the year 1840, and was of very great magnitude. It has been described in considerable detail by Dr. Coan. And in the year following the outbreak it was visited by Lieutenant Wilkes and investigated with considerable care. It first manifested itself in an open crack, about 7 miles to the eastward of Kilauea, where a small quantity of lava was disgorged. At various points still further eastward lava came up in small quantities through fissures. The main eruption broke out about 12 miles from the sea-coast and about 25 miles east of Kilauea, and an enormous mass of lava was outpoured. It spread out in a stream nearly 3 miles in width, and reached the ocean at Nanawale. This eruption has often been spoken of as emanating from Kilauea and flowing underground to the principal point of outbreak. This view is scarcely credible. It is much more rational to suppose that the several points of outbreak had a common connection with a single line of fissure, and that the principal outbreak took place merely at the lowest point.

The number of circular pits in the vicinity of Kilauea and along this line of volcanic energy is considerable. Fifteen of them are now known, and it is by no means certain that the number is not considerably greater. They are situated in a region which is densely forest-clad, and which the traveler crosses only by means of trails hewn through the woods. On the road from Kilauea to Puna I passed within a dozen yards of a large one without seeing it until I was called back by one of my packers, who pulled aside the dense curtain of shrubbery, disclosing a pit nearly 2,000 feet wide and 600 feet in depth. These pits are either circular or elliptical in form, with remarkably regular outlines, and the sides are partly precipitous, and for the remainder, very steeply sloped. They are usually spoken of as of craters, but as the term crater in this connection seems to commit us to the idea that they have some affinity with true craters, I avoid using it.

Along the line of volcanic activity, extending from Kilauea to Nanawale, are numerous cinder cones. There are several lines of these, which are very nearly parallel and separated by short intervals. Most of them are quite small, but there are two or three of considerable magnitude. The larger ones are perfectly normal in structure, and are composed of lapilli of ordinary type. I have already remarked upon the absence of normal cinder cones upon the mass of Mauna Loa, and so far as Kilauea is concerned, the number of such normal cones is small, considering the mass of mountain.

It is difficult at the present time to distinguish the character of the many eruptions which have no doubt taken place along this line of activity. Vegetation is so luxuriant, the rainfall is so great, and the disintegration of the rocks by atmospheric weathering is so rapid, that eruptions a few centuries old are completely obscured as to their details. But at many points we become conscious of the fact that many large eruptions which seem to emanate from this line of volcanic activity have followed each other in rather rapid succession. If we go from Kilauea to the sea-coast on the southeastern side of the island, then follow this sea-coast for 20 or 25 miles to the easternmost cape of Hawaii, we shall cross many large streams of lava flowing directly seawards from the principal line of fissure five miles inland. Many of these streams appear to be moderately recent; that is to say, not more than a few centuries old. No soil as yet covers them; still there are numberless plants which grow upon the rocks themselves as readily as the ivy grows upon the walls of an old cathedral; and lava beds without a trace of soil are completely overgrown with luxuriant vegetation.

We are now prepared to conjecture something concerning the origin and mode of formation of this great depression of Kilauea, though our conclusions must of necessity be very limited. There can be very little doubt that its formation has been gradual. It is quite possible that violent action has in past times occurred, though there is no reason to suppose that it has ever been of such extreme character as would be necessary to produce so large a cavity at a single convulsion. The long continuance of just such a process as we now either perceive clearly or directly infer would be ample to explain its development. The many pits in the neighborhood, and of which Poli-o-keawe is a good type, may suggest to us the condition of Kilauea in its earlier stages. We may conceive a column of lava penetrating the rocks upward until it nearly or quite reaches the surface. At times it may erupt and pour out lava streams of variable magnitude. Once established, this column remains for a long time in a liquid condition, being kept hot by the rising of intensely heated gases from below or by convection currents which these rising gases set up. The height of the column oscillates from month to month or from year to year, but within small limits—say a very few hundred feet of altitude. Gradually the rocks in its vicinity are melted, but at a slow rate. This rate would depend upon the difference between the supply of heat from below and the loss of heat by escape into the atmosphere. The melting of the rocks in contact with the lava column not only enlarges the pipe itself but robs the superficial beds adjacent to the orifice of their support, and they sink down in successive spalls and are gradually remelted below. From time to time the reservoir thus formed is emptied temporarily by some eruption from a distant fissure or crater with which it is for a time brought into connection. But this distant eruption ceases, the fissures and subterranean conduits close up, the reservoir is refilled, and the lava column comes back to its former condition, and the process is renewed. That the caldera has been enlarged by the sinkage of large fragments is sufficiently borne out by the appearance of such fragments, more or less sunken, all around its rim. At nearly every sharp shock of earthquake the slipping of these fragments is witnessed, and I noted in the great wall which is on the left as we descend by the trail from the Volcano House into the caldera the fresh face of the escarpment between two parallel lines which appeared to mark the former height of the sunken spall in front of it. Such marks would doubtless appear more frequently were not the growth of vegetation so rapid and the rains so incessant that they are soon obliterated. On the northwestern verge of the pit a large spall was shaken down by an earthquake in 1878, and the manager of the Volcano House assured me that such occurrences had happened under his observations many times. This is understood to be the case by those who have frequent occasion to pass by Kilauea.

The height of the lava column in Kilauea has varied considerably from year to year. It is probably higher now than at any time within the historic period. When Ellis visited it in 1823 it must have been nearly four hundred feet lower. By its rising it has filled up completely the great inner cavity within the historic black ledge, and has built up the whole floor of the caldera with new lavas to a height which is variable, but which may average nearly a hundred feet above the former level of the ledge. The rise of the lava column, however, has not been uniform, but the height has oscillated up and down with a gain of altitude on the whole of about 400 feet in sixty years.

Beneath the floor of the caldera we may conjecture the existence of a lake of far greater proportions than those which now expose a fiery surface to the sky. The visible lakes might be compared to the air holes in the surface of a frozen pond. That such a lake of large proportions really exists with a thin covering of congealed lava, and open in only two spots, may derive support from the fact that whenever eruptions take place within the caldera (and no year passes without several of them), they do not come from the overflow of the open lava pools, but break out anywhere in the floor, and always at unexpected places. Wisps of steam are seen rising at many points, and even from cracks in the summit above. A few feet and often a few inches below the surface within these cracks the temperature is scalding hot, and sometimes hot enough to char a stick thrust into them. It is very doubtful, however, if this great lava lake underlies the whole caldera and the cliffs of the entire surrounding rim. This is matter of the purest conjecture without facts sufficient for guidance. We can perhaps justify the belief that the open surface of liquid lava is but a fraction—perhaps a small fraction—of its area, but we have no means of estimating either its extent or configuration.

It may be asked, Why should not the solid crust above this lake break up and sink like the crust in the open lakes? This may be readily answered. The rocks forming the floor of the caldera and of the surrounding platform above it are pahoehoe, which is very porous and spongy. A glance at a fragment of it shows thousands of closed vesicles. Its specific gravity, then, is small and much below the absolute density of the constituent material. In a word, it is so light that it can readily float. The scum formed on the burning lake, though it may have such vesicles at first, yet by remaining a long time in contact with hot lava in a viscous condition the vesicles disappear just as they do from glass in the glass furnace by protracted heating, and the specific gravity of the scum becomes about equal to its absolute density, until at last it is heavy enough to sink. That it is very viscous at the moment of break-up is seen distinctly in the easy way the cakes bend down their edges like a piece of leather when they sink.

It may be interesting to compare Kilauea with other depressions around volcanic vents, which seem to show similar action. Perhaps the most striking instance outside of these islands is the summit of Teneriffe and the great "*cirque*" around the Pico de Teyde. This has been splendidly figured by Fritsch, Hartung, and Reiss, and shows a nearly elliptical depression in the summit portion of the island, whose diameters are about twelve and eight miles respectively, while the engirdling walls are from one thousand to fifteen hundred feet high.* In the center of it rises the lofty cone Pico de Teyde, which is frequently steaming, and has within the last hundred years given forth lava. The peak rises more than four thousand feet above the floor of the cirque, and nearly covers it with its ample base. The descriptions and illustrations of it suggest an origin for the depression similar to that propounded here for Kilauea. The central cone may be in chief part of subsequent accumulation.

Another instance apparently similar is the Volcan de Taal on the island of Luzon, about forty miles south of Manila.

---

* This is larger than Haleakala, though the depth is considerably less; but the indications are that much of the depth has been refilled by materials ejected from the central and surrounding orifices.

CHAPTER V.

MAUNA LOA.

Our next objective point is Mauna Loa. From almost every point on the rim of Kilauea the great dome of this mountain is in full view. In truth, it is difficult to imagine a much more advantageous way of viewing so vast a mass than from the rising knoll of Kilauea. No forest obstructs the view, and all that portion of the dome which lies higher than the 4,000-foot level is fully presented. We cannot realize its magnitude. It rises nearly 10,000 feet above us, and occupies about 130 degrees of the horizon. Its slopes are very gentle and uniform, and the absence of details is its most peculiar characteristic. There are no ravines, no spurs nor sharp crests, no knobs, cinder cones, nor monticules. In one or two places, however, may be seen a few minute prominences, which we recognize at once as the abortive attempts to form cinder cones, which characterize many of the eruptions of this volcano. The roundness and smoothness of the pile is, in a topographical sense, its most conspicuous feature. The mottling of light and shadow upon its surface gives us some premonition of its real character. Here a long dark streak, reaching many miles down its slope, indicates some lava-flow of recent date, which has not, as yet, lost its blackness by weathering. Far up toward the summit these black streaks increase in number and merge together. The upper limit of vegetation is generally well marked, but forms a very uneven line, here ascending, there descending. The greater part of the altitude of the mountain seems to be above this limit.

From the Volcano House there are several methods of approaching the mountain. Perhaps the most common one is by the trail leading back to Kapapala, over which we came as we approached Kilauea from Kau. Having traversed this route once, I determined to select one leading more directly to the object. Leaving the Volcano House, and passing along the northern rim, we moved away from Kilauea towards the northwest, crossing the depression or saddle which separates the two mountains. The trail soon enters a region covered with patches of koa forest and a long rich grass strikingly similar to the grasses found high up in the mountains and plateaus of the Rocky Mountain region. The koa forests are eminently characteristic of the island, and form one of the most attractive features of its vegetation. The trees are large, having trunks often two and a half to three feet in diameter, and growing very closely together. The wood of this tree is very dark and hard, having a color somewhat similar to the black walnut, and even more ornamental. It is also quite as hard, if not harder. It was from these trees that the natives formerly made their canoes. The koa does not flourish on this island below 4,000 feet, though straggling groves and individual trees are occasionally found at considerably lower levels. From 4,000 to 6,000 feet, and sometimes higher up, it is the dominant forest tree. The koa forests, though still rather extensive, were formerly much more so. But they have within the present century been greatly ravaged by wild cattle. The alternation of forest and grassy park is very pleasing, and if water were abundant the region would be a paradise.

The trail soon enters upon the naked lava. Wide fields of pahoehoe stretch illimitably to the south and southwest, alternating here and there with bands of clinkers. About four miles from Kilauea we reach the foot of the ascending slope of Mauna Loa, and for a time we leave the naked lava and enter upon a rising slope of soil, clothed with grass nearly waist high. The ascent now becomes somewhat rapid, though never steep. After gaining about a thousand feet of altitude, we reach an abandoned ranch named Ohaikea. Here stands a decaying and abandoned cottage built out of planks sawed by hand from koa trees. Several cisterns or large hogsheads contain rain-water gathered from the roof. The mountain slopes in the vicinity are deeply clothed with soil, and appear to have enjoyed for a long period immunity from the devastations of flowing lava. During the summer season the climate here is rather dry, but during the winter the rainfall is copious. The altitude is about 4,800 feet.

Leaving Ohaikea, we turn sharply to the southwestward, moving along the flank of the mountain, and, on the whole, gradually descending. A few miles in front of us is a long lava stream which flowed in the latter part of 1880, forming what is termed the Kau branch of the great eruption of that year. It ends just where the slope of Mauna Loa meets the slope of Kilauea. As it consists mostly of aa, and has a width of about half a mile at its narrowest, it is better to go around it rather than to attempt to cross it. After a ride of about four miles from Ohaikea, we reach the termination of this flow, and find it ending in a bristling mass of angular fragments, fringed with a steep slope about 40 or 50 feet in height, and repulsive to the last degree. The whole mass seems to have thoroughly cooled, and no traces of heat were detected. Skirting around the end of this flow, we again move a little way up the slope of Mauna Loa and thence along its base. Upon our right is a steep slope, suggestive of the face of a terrace, rising about 300 feet very abruptly. Whether this is another and higher member of the series of terraces which we saw in the vicinity of Hilea, Pahala, and Kapapala it is difficult to say, but there are many appearances which suggest that interpretation. It is tolerably well marked for the distance of about 16 miles upon the eastern flank of the mountain, and at either end is overflowed and buried by numerous eruptions. Traversing the base of this terrace seven or eight miles, we at length ascend to the summit of it. Winding among the koa groves, and through the open meadows of mountain grass, we at length reach an inhabited ranch called Ainapo. Here a large cistern has been constructed of cement and filled with rain-water from the roofs of the houses.

Ainapo is a charming spot in the summer time. It is situated at an altitude of about 4,200 feet, among open grassy parks and groves of koa. The air is cool, and the trade-wind ever blows gently. The view seawards is a commanding one, though often obstructed by the drifting banks of trade-wind clouds. Far down the mountain slopes, which here descend with a gentle declivity, may be seen the broad, black, desolate fields of pahoehoe which have flowed from Kilauea over a very gently descending plain to the southwestward. Still further beyond is the faint glimmering expanse of the sea merging into the sky without a visible horizon. To the eastward is seen the cloud which always overhangs the furnace of Kilauea, and when the night comes the glare of its fires suffuses the clouds with a rosy red "flaring like a dreary dawn." The rim of its amphitheatre is distinctly visible, and just behind it the chaos of jagged rocks which encircles the fiery pool of Halemaumau.

Ainapo is the last halting place in the ascent to Mauna Loa where the traveler will find the three great requisites of camp life, fuel, water, and grass, in proximity to one another. Using this as a base station, my first objective point was the sources of the last great eruption of Mauna Loa, 1880–'81. They are situated high up on the northeastern flank of the mountain. To find them a guide is necessary. A determined traveler might reach them without one, but only with immense labor and loss of time, if not with loss of animals. I had been so fortunate as to secure the services of a native goat hunter, who knows the entire southern portion of Mauna Loa more thoroughly than any other man. Leaving Ainapo early in the morning, we moved obliquely to the northeastward, slowly and steadily gaining in altitude. For about 7 miles we followed cattle trails leading through koa groves and grassy parks; then crossing a stream of aa we turned sharply up the mountain slopes, and in an hour or two had reached the upper limit of vegetation at an altitude of about 6,700 feet. Hard by was the Kau branch of the flow of 1880. Here we camped for the night, the afternoon being spent in foot excursions among the various lava streams, most of which are of very recent origin, though of unknown dates. These lavas vary but little in composition, all of them being highly olivinitic and heavily charged with iron. They are very vesicular, and really compact specimens are difficult to find. Most of the lava has taken the form of pahoehoe, though many broad fields of aa alternate with it.

At the first streak of dawn we began the final march to the summit, or rather to a point a few miles northeast of it. The whole distance lies through barren fields of naked lava. Innumerable streams descending from the summit are all around us, showing by the varying degrees to which they have been affected by weathering just so many different epochs of eruption. These streams vary in width from a few hundred yards to a mile or more, and many of them have taken the form of aa. Between them are broad lanes of pahoehoe, along which we lay our line of travel. Frequently the streams of aa are seen to coalesce further up the mountain, and whenever this is noticed it becomes necessary to look for an easy place to cross the aa on the right or left, in order to find another lane of pahoehoe which will carry us further up. The crossing of one of these clinker fields is harassing to the animals. So keen and sharp is the loose rubble of clinkers that the fetlocks of the mules are lacerated as they flounder through it. The poor brutes dread the ordeal not a little, and it usually requires something more than mere moral suasion to induce them to enter upon it.

The lavas present many curious forms along the way, arising from the varying circumstances under which they have cooled and solidified. Wherever it has poured over sharp ledges or down steep slopes it takes the form of a matted and tangled mass of large ropes. On the gentler declivities it is, for the most part, typical pahoehoe, spreading out in large belches 20 to 40 feet square with wrinkled surfaces. In a few places we found some appearances which were difficult to explain. Long straight gutters or trenches looking like ditches in the soil which have been dug by hand, are seen with banks a foot or a foot and a half high, and with the bottoms two or three feet wide. Within these troughs the lava reproduces in stone the appearance of water running swiftly through a wooden flume.

As we approach the summit the fields of clinkers become more abundant and of larger proportions. The upper 2,000 feet of the mountain seems to be surfaced chiefly with this form of lava. The guide showed admirable acuteness in dodging the greater part of these formidable fields and selecting lanes of pahoehoe. But at the very best, we must have crossed in the ascent rather more than two miles of aa, and the condition of the animals was most pitiable.

At an altitude of 11,800 feet we were obliged to dismount and tether the mules to the rocks. For in front of us and on every hand was a chaos of clinkers and rubble, with yawning cracks and fissures, which must be crossed on foot. Lashing our instruments and photographic camera on our backs, we journeyed about a mile, crossing several large fissures until at last we reached one which gave unmistakable signs of being the object sought. It stretched as far as the eye could follow it both up and down the mountain. It was very narrow for the most part, but at several places it expanded into large holes where no bottom could be seen and from which there still issued copious volumes of steam. From these great holes the last drainings of the lava flow are seen streaming away down the mountain slope. It has the appearance of basaltic obsidian. It is highly vesicular, and the vesicles, contrary to the usual habit in basalt, are very elongated and much drawn out, like those so often seen in rhyolite or pumice. It has a dark olive-green color, and as we step upon it it grinds, shatters, and crackles beneath our feet. Nothing like a cone has been erected over the orifices. The uppermost of these large vent holes gave rise to a stream of lava of enormous dimensions which flowed more than half a mile wide down the northern slopes of the mountain, reaching nearly to the base of Mauna Kea and spreading out on the intervening plain between the two great volcanoes, in a field the extent of which could only be guessed, but which from an eye-estimate seemed to be 12 or 15 square miles in area. This constituted the first part of the eruption and flowed for about three weeks. Soon afterward another vent opened along the same line of fissure about a mile further down the mountain slope, and gave issue to the Kau branch of the flow, which has a length of about 10 miles and a width varying from half a mile to a mile and a quarter. Still again along the same line of fissure and about half a mile further down the slope broke out the principal branch of the eruption which flowed down the northern side of the mountain into the intervale between Mauna Loa and Mauna Kea, then deflected to the right and flowed eastward to within half a mile of the town of Hilo. The length of this branch of the flow is about 45 miles and the width varies from half a mile to 3 miles. No cone was formed at any of the three upper orifices. Many parallel cracks are seen on either side of the main line of fissure, from some of which lava issued at the same time. In the vicinity of the upper orifices the cracks widened out and gave vent to the lavas along a considerable line of opening on either side of the hole.

To convey a more complete idea of the circumstances connected with this eruption, it seems necessary to explain one of the features connected with the form and arrangement of the mass of Mauna Loa. The base of the pile is by no means of circular form, but its figure is elongated into an oval shape, with the major axis extending northeast and southwest. In the elongation towards the northeast we may conceive some analogy to a flat and feebly-pronounced spur, reaching the sea in the vicinity of Hilo. Along this spur the slope of the mountain is less than along any other descending element of its cone. Nowhere, however, does it present that sharp, ridgy appearance which we usually associate with mountain spurs; but the projection is well rounded, and when viewed as a whole is remarkably smooth. No ravines and gorges, however small, occur. It is along the line of longest slope and minimum declivity that the rupture took place. The cracks extend directly from the summit downwards and are many in number, always preserving a very approximate parallelism. The appearances indicate that the lavas broke forth from more than one of these fissures and from very many points, the numerous streams finally coalescing into three principal ones. There is, however, much difficulty in tracing out all the subsidiary facts of the case, for there is nothing to guide us here excepting the recent aspect of the several constituent streams. It is not a little remarkable that an event of such importance has left none of the ordinary monuments of a great eruption. Of the stupendous mass of material extravasated only small traces have been left around the sources. All flowed down the mountain sides and came to rest in localities many miles away. The vents from which the greater part of the lava issued were found to be much smaller than was anticipated. It seems extraordinary that so vast a mass of lavas could have issued from orifices so small; but, on the other hand, the flows were continuous for a period of about eleven months, and when the vents were in action the lava was ejected with immense velocity.

It is difficult to obtain from eye witnesses of this eruption any very detailed and precise account of the particular phenomena which we would most like to have a full description of. It was seen by several persons who viewed it from the base of Mauna Kea, which on the whole would afford about the best standpoint for a distant observation. They all speak of seeing fountains of fire projected to a great altitude, variously estimated at 400, 500, and even 1,000 feet. These, however, must be regarded as mere guesses. No one could see whether this fire consisted of jets of lava or of incandescent vapors. All of the observers were impressed with the idea that the lava itself was projected upwards in fountains to a great altitude. Of the varying aspects of the eruption during its various stages no circumstantial account is obtainable.

A very interesting circumstance connected with this eruption was the fact that for several months before it took place an intense activity prevailed in the great summit caldera. As soon as the eruption broke out, the fires on the summit appear to have ceased suddenly, and from that day to this have shown no indications of renewal. It seems little doubtful that the eruption tapped the lava lake at a lower level. The highest vent of the eruption of 1880 is situated about 800 feet lower down than the surface of the lavas in Mokuaweoweo. Nor is this coincidence by any means unprecedented, for according to all accounts the same thing happened in 1855 and again in 1859, the great eruptions of those two years being preceded by intense activity at the summit followed by complete but temporary repose. With none of the last four great eruptions of Mauna Loa has Kilauea shown any apparent sympathy.

There is a strange fascination in wandering over this vast expanse of desolation. No doubt the dominant idea is the immensity of it. The best conception of the magnitude of Mauna Loa is to be obtained by attempting to traverse any limited district of it on foot. Mile after mile may be traversed but the landscape seems ever the same. All the great landmarks seem to stand just where they stood an hour before. One locality is as much like another as any two parts of the Iowa prairies or of the Great Plains of Colorado. The only alternations are from *aa* to pahoehoe. Traces of recent eruptions are seen everywhere, but all the views are fragmentary. So extensive has each and every one of them been that the greater portions of them always reach far beyond the limits of vision, and mingling together are lost in the confusion of multitude. The imagination is discouraged at the thought that this colossal pile has been built up by thousands upon thousands of these eruptions.

In the vicinity of the last eruption stand several small knobs or cones of unknown date, which we may examine in detail. Their origin is no doubt the same as that of ordinary cinder cones. In the truncated summit of each is a crateriform depression which is, in fact, a true crater. The pile is built up of large clots and fragments of basalt thoroughly welded together, as if the constituent fragments had fallen in a pasty condition. The fine lapilli and volcanic ashes seen in ordinary cinder cones are wanting here. They are insignificant in size, none of them exceeding 120 feet in height. One of them has been split asunder by a large fissure traversing it, and a little lower down on this same fissure is situated one of the vents of the last eruption.

The source of the great eruption of 1855 is situated not far from that of 1880, lying a little further to the northwestward, a little below the line where the great mountain spur begins to descend to the northward. I could not feel quite sure that I had hit upon the exact locality though I considered it probable that it was found. The most remarkable circumstances connected with it are the comparatively light traces which have been left of an event so momentous, and the indistinct character of the few which remain. This outbreak flowed unceasingly for thirteen months, and the lava covered an area of nearly two hundred square miles; and yet at the fountain head the earth shows but a few insignificant wounds. This eruption also issued from a crack radiating from the summit of the mountain along the main spur. The general appearance of the ground in the vicinity is very similar to that presented in the eruption of two years ago. Of the great flows of 1855 and 1881 we shall see more during a later stage of our journey.

Descending the mountain slope to reach our camp at the timber line, we have many opportunities to note at our leisure the varying phases of the flowing lava streams which are still perfectly preserved, even in those eruptions which have considerable antiquity. It has been repeatedly noted hitherto that the solidifying lava takes the form of either pahoehoe or aa. We have abundant opportunities to verify the conclusion already reached that it takes the form of aa wherever the cooling mass is very large or where a considerable body of hot and highly viscous lava moves at a very slow rate. The pahoehoe, on the other hand, is formed by highly liquid belches shot out in comparatively small quantities from the interior and hotter parts of the *coulee*. Nearly every stream, and perhaps literally all of them, show in some parts of their courses long tunnels, and in numerous places the arch above the hollow pipe has fallen in, leaving a deep pit in the ground and revealing the continuation of the gallery both upwards and downwards.

Some of these galleries are, no doubt, miles in length, and I have been informed by credible persons that they have followed old galleries for three or four miles without finding any opening or termination. Their dimensions vary considerably, being sometimes sixty or eighty feet in diameter, and frequently constricted to eight or ten feet. The number of these tunnels in the mass of Mauna Loa must be vast indeed. As a general rule, they are not disclosed, and we become aware of their presence only when we find a spot where the arch has fallen in. Undoubtedly these tunnels are the agency by which the lavas are able to flow to such immense distances as forty or fifty miles from the vent and yet preserve to the end a high degree of liquidity. The radiating power of these lavas is very great, but their conducting power is excessively feeble. Wherever a fresh molten surface is exposed nakedly to the atmosphere, it blackens with very great rapidity, and a solid skin is formed in a very few minutes. But this skin immediately acts as a protecting mantle or jacket to the hot lava within, and its very feeble conductivity preserves the interior heat for a long time.

Reaching our camp at nightfall, we passed a comfortable night and the next morning returned to Ainapo. Halting a day to recuperate the weary and half-starved animals, we began the ascent to the summit platform of the mountain. For this undertaking a guide is very necessary. There is one, and only one, route by which it can be accomplished with comparative ease. To deviate from that route is to become entangled in a wilderness of impassable clinker-fields, where progress is possible only on foot, and even then with extreme difficulty. My guide knew the trail thoroughly, and is perhaps the only guide now living who does know it. By the proper route it is possible to ride to the apex of the mountain without dismounting from the saddle. As a feat of mountaineering it is not worth mentioning. The only difficult feature about it is the very great length of the journey, for, as the trail runs, the distance to the summit is fully twenty miles from Ainapo and the altitude to be gained is about 9,500 feet. It was thought best, therefore, to make two journeys of it, the first day being consumed in traveling to the upper limit of vegetation. The trail winds among groves of koa and grassy parks, here leading over a flat plain, there ascending a strong acclivity. Upon our left is the border of a compact forest, which covers the southern slope of Mauna Loa between the altitudes of 2,000 and 6,000 feet. To the right of us the country is much more open, and the trees stand in clusters. This dissimilarity has its origin in a difference of rainfall. The density of the forest is everywhere proportional to the amount of precipitation. Concerning the climatology of this island, I shall perhaps have something to say hereafter.

As the distance to be made is only about seven miles, it was thought a good opportunity to indulge in a little hunting. Game is perhaps as abundant here as in any locality in the world, but the kind of game and the kind of hunting are certainly exceptional. The mountain literally swarms with pigs, cattle, and goats, all of which are thoroughly wild. Wild horses and asses are also met with, and occasionally even wild mules. The commonest beast is the hog, and the facility with which it was caught and slaughtered by my native guide was most amusing. A large well-trained dog suddenly darts into the shrubbery, and in a moment the squealing begins. The guide, dismounting from his horse, at once comes to the rescue, and as the pig and the dog are tussling in opposite directions he seizes the former by the hind legs, trips his fore legs, turns him upon his back, and with a large sheath-knife severs the jugular vein in less time than it takes to tell it. The dog and the kanaka are then after another pig and with the same result. Many of these animals are vicious creatures, with sharp tusks four or five inches in length, and an assault upon them is apt to be dangerous, especially to the dogs. Most of them are black, with singularly long snouts and large heads. These pigs were abundant in the islands at the time of their discovery. They are said to have a very close resemblance to the wild hogs which are so abundant in the East Indian archipelago, from which part of the world they were no doubt derived.

Wild goats are also very abundant. These were brought to the islands near the close of the last century by Vancouver, and they have multiplied rapidly. The native method of hunting them is decidedly unique. The goat hunter follows a flock on foot. As he approaches they gallop away over the rocks, leaving the pursuer far behind. But they soon halt, tired and blown by their exertions, while the kanaka keeps on. It becomes a question of endurance between the steady jog-trot of the pursuer and the alternate halts and spasmodic efforts of the pursued. The kanaka wins every time. In the course of a couple of hours the animals are too weary and too much discouraged to flee further. Reaching the first laggard, the hunter breaks its hind legs across his knee, and the remainder of the flock are treated in like manner. Returning upon his track, he skins the animals at his leisure. The goats have little or no value except for their skins, many thousands of which are exported from the islands annually. The flesh of the ewe goat proved to be much better meat than I had ever supposed, being quite as good as veal or mutton, with a flavor intermediate between the two.

The wild cattle are also very abundant. They were introduced by Vancouver in the year 1792, and a strict *tabu* was laid upon them for many years. They multiplied with great rapidity, and in the year 1825 large herds of them were running wild over the mountains of Hawaii. They are very destructive to the forest vegetation, and efforts have been made from time to time to check their increase, but no check has been found possible beyond the steady decimation by cattle hunters and the limit of the amount of suitable food. They have for many years been hunted and slaughtered in large numbers for their hides. Within the past few years the development of the sugar plantations, with the increased number of laborers employed, has created a demand

PANORAMA

U. S. GEOLOGICAL SURVEY

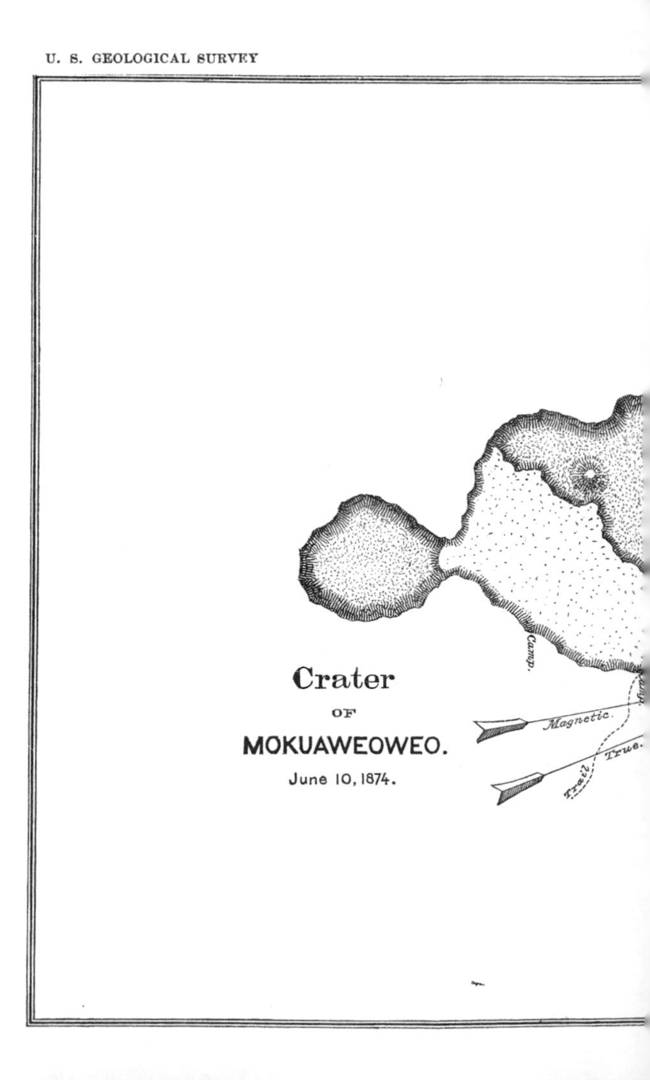

UAWEOWEO.

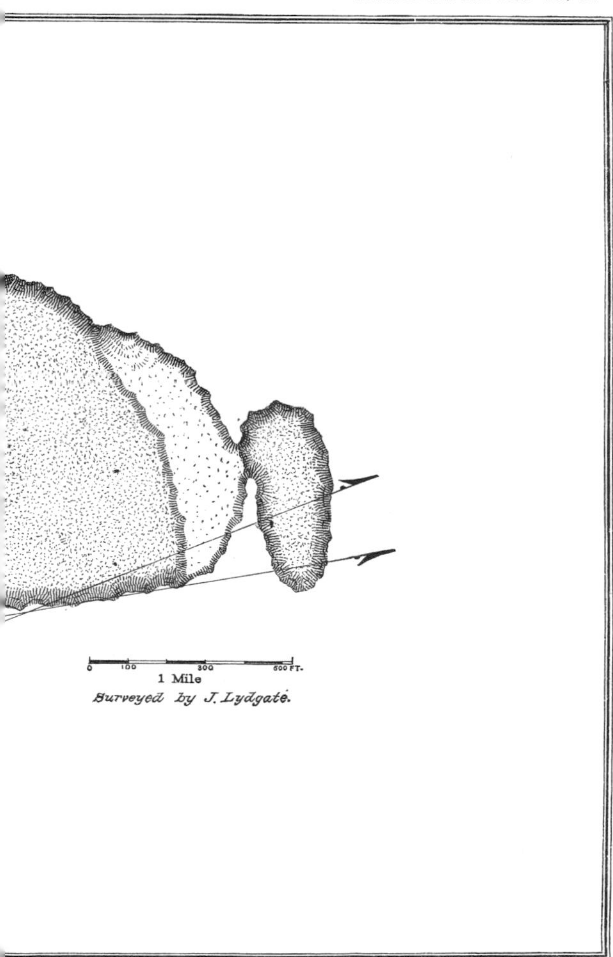

for the dried or jerked beef, which is brought down from the mountains by the hunters, though not in great quantities. From all accounts it would appear that the koa forests have been very much ravaged by these animals, which not only devour the young seedlings, but strip the bark from the older trees. The animals are very timorous in the presence of man, for they have been hunted for several generations and slaughtered without remorse. The sense and fear of danger thus acquired have made them very cunning. Their favorite places of retreat are the dense jungles, where a horse and rider cannot penetrate. In seasons of drought they are obliged to leave their fastnesses to seek for water, and they make their journeys always in the night-time, returning to their retreats before daylight. They have the faculty of noting during the dry weather the fall of a distant shower in the day-time, and when the night comes they make a long journey to the locality with an unerring instinct, and seek the pools of water in the hollows of the rocks. The native hunter is still more cunning than the brute, and spends the night upon the trail which he anticipates they will follow in pursuit of water, and shoots them down as they pass.

We reached the upper limit of vegetation early in the day and spent the afternoon in hunting pigs and wild cattle. So far as the pigs were concerned we could never go amiss. The country swarmed with them, and at every hundred yards or so a snorting porker would scud out of the path. We were also fortunate enough to secure a fine large cow, which furnished us with a supply of beef for several days. It seemed at first like utter ruthlessness to destroy so noble an animal merely to get two or three beefsteaks, but circumstances alter cases.

Our camp was located at about 6,700 feet above sea-level, where vegetation was still tolerably abundant, though it ends abruptly a little above us. The outlook down the mountain slopes was much obscured by clouds, most of which floated below us, while now and then a bank of mist enveloped us. Whenever the curtain of cloud and fog drifted away we could look down upon the broad plain of Kilauea and see with perfect distinctness the outlines of the great caldera and the rocky pile around Halemaumau. At night the glare of its fires tinged the clouds with vermilion, which came and went as the incandescence of the lavas beneath fluctuated.

At the first streak of dawn we began the final march to the summit. In half an hour the last trace of vegetation was below us, and the track lay over fields of pahoehoe ever mounting upwards. Nowhere is a trace of soil to be seen. The lava lies just as it was originally poured out and cooled. Here and there a stream of lava has been considerably weathered, its original black having changed to a rusty brown or reddish color, but nowhere has it broken down into soil. Nowhere is the travel attended with any great difficulty other than such as attends the prolonged exertion of mounting a constant succession of steep acclivities. Here and there are faint marks of a trail made by preceding travelers to the summit, but ordinarily no trail is visible. The guide showed all the tact and instinct of an Indian in finding his way. On either side are clinker fields sending down their long fringes and fingers, and a keen memory must have been necessary to enable him to recall the exact lane which he must follow to escape an entanglement in such a dreadful net. But no error was committed. After a journey of about five hours without a halt, half of which was accomplished on foot, leading the animals to save their strength, we reached the summit platform. One animal showed signs of distress from the thinness of the air at an altitude of about 11,000 feet. We left him tied to a pinnacle of rock to await our return.

The summit of Mauna Loa is a broad and large platform about five miles in length and four miles in width, within which is sunken the great caldera called Mokuaweoweo. The distance from the point where we first reach the summit to the brink of the pit is about a mile and a half. The surface of the platform is much more rugged than the slopes just ascended. It is riven with cracks and small faults, and piles of shattered rock are seen on every hand. Nowhere is there to be seen the semblance of a cinder cone. Doubtless many eruptions have broken forth from the various fissures on this summit, but only here and there can insignificant traces of such catastrophes be definitely distinguished. The absence of fragmental ejecta is extraordinary. The shattered blocks, slabs, and spalls which everywhere cumber the surface appear to have resulted from the spontaneous shivering and shattering of the lava sheets by their own internal tensions as they cooled.

We come upon the brink of the caldera very suddenly. The cliff beneath is quite vertical, descending at the point where we first reach the rim a little more than 600 feet. The spectacle is a very imposing one, far more so than the view of Kilauea from the rim of its amphitheater. The horizontal dimensions of Makuaweoweo are somewhat less than those of Kilauea, but the depth is much greater and the encircling walls are even more precipitous and continuous. The only break in the walls where a descent is possible is found upon the northeastern border. The floor beneath has the same desolate appearance as that of Kilauea, being covered with numberless surges and slops of lava which have poured over it and congealed, remaining to-day just as they were when they first cooled. In the central part is a ledge surrounding a still lower depth, corresponding no doubt to the historic black ledge of Kilauea which existed fifty years ago. The depth of this ledge is perhaps 175 or 180 feet, and its position is indicated on the accompanying map.

At the time of my visit there was no volcanic action whatever. Not even a wisp of steam could be detected issuing from any point. The lava lake within the ledge was completely frozen over, looking as solid and ugly as the rocks on which we stood. In truth, the strongest feeling impressed upon the mind was that of superlative calm, solitude and desolation. And yet, scarcely two years have passed since these cyclopean walls witnessed displays of volcanic energy which are probably unrivaled in beauty and magnificence. Within the lower ledge was formerly a large lake of liquid lava. Of its dimensions we have no account further than that it much surpassed in extent either of the lakes of Kilauea. At brief intervals from out of the surging mass liquid columns of fire were shot up with fierce explosions to the height of 500 or 600 feet, falling back again in a prodigious spray of glowing masses into the pool from which they ascended. This scene has been described to me by Mr. H. M. Whitney, who visited it in the year 1878, and who, standing upon the brink, marked the particular, well defined stratum of red lava in the opposite wall against which the summit of the fountain was ranged in the line of vision. This stratum is clear and unmistakable, and its height above the surface of the frozen lake is between 500 and 600 feet, and nearer the latter than the former figure. Only a few days before the last great eruption of 1880–'81 the glare of the fires within this lake and the spasmodic bursts of flame from its spouting columns had been seen from many points illuminating the clouds above, and on many occasions for a year previous the same phenomena had been witnessed. But no trace of illumination has been seen since the eruption. So infrequent are visits to the summit that I have not been able to learn of any traveler who has reached it within the last two years, and it is impossible to conjecture what took place while the eruption below was in progress. Nor is this the only instance in which the great eruptions of Mauna Loa have been preceded by intense activity and followed by perfect quiescence at the summit. According to all accounts the same thing happened in 1855–'59. There seems to be good reason for presuming that an eruption upon the flanks of Mauna Loa is so connected with the summit orifice as to exhaust for a time the volcanic energy which had previously been accumulating there. The period of repose at the summit after any eruption is very uncertain, and want of accurate observation does not enable us to draw any specific conclusions in this matter. Stated, however, in the most general way, it would seem that within the mountain there is continuously in progress an accumulation of eruptive energy and materials, which at first seek an outlet through the summit orifice; but as the accumulation goes on the mountain itself is ruptured. Lower down upon its flanks the lavas are discharged and the volcanic energy is for the time being depleted. The periods between eruptions, however, are quite irregular, varying from four to ten or eleven years.

The length of the main caldera is a little less than three miles and its width about a mile and three-quarters. Its floor, viewed from above, appears to be composed of a series of flat surfaces occupying two distinct levels, the higher upon the surface of the black ledge, the lower lying within the ledge. Upon the western side is a small cinder cone standing close upon the border of the black ledge. It is the only cone visible, either within the caldera or upon the surrounding summit. Its height is about 125 or 130 feet. It was seen in operation, throwing up steam, clots of lava, and lapilli, in the year 1878. The only other diversifications of the floor are many cracks which traverse it, the larger of which are distinctly visible from above. Some of them are considerably faulted. There is no difficulty in recognizing the fact that the whole floor has been produced by the sinkage of the lava beds which once continued over the entire extent of the depression, their under sides having been melted off most probably by the fires beneath. The lava beds in the immediate vicinity of the brink upon the summit platform wear the aspect of some antiquity. They have become brown and carious by weathering, and, although no soil is generated, little drifts of gravel are seen here and there mixed with pumice. Since the caldera was formed there is no indication that the lavas have anywhere overflowed its rim. And yet it is a very strange fact that within a half mile, and again within a mile to a mile and a half, lavas have been repeatedly erupted within the last forty years from the summit platform, and have outflowed at points situated from 700 to 900 feet above the level of the lava lake within. Traces may also be seen, at varying distances back of the rim, of very many eruptions in which the rocks betoken great recency, although no dates can be assigned to their occurrence.

The descent into the caldera is accomplished with no great difficulty at the northeastern extremity. Moving around the brink to that end we find just beyond it a smaller pit separated from the main caldera by a narrow ridge. The wall here is at its lowest point and is also broken down into a rubble of large blocks and rugged fragments, where we may carefully pick our way downwards. Reaching the summit of the black ledge, immediately above the platform of the frozen lava lake, considerable search is necessary to find a practicable spot, where we may descend to the lower level. The ledge is exceedingly sharp and, for the most part, vertical, though in several places it looks practicable. The easiest point is found at the foot of the main western wall, where a talus of great bowlders enables us to find a rugged path to the bottom. It is not without some sense of apprehension that we venture to tread upon what is literally, the crust of a sleeping volcano, and that one the mightiest volcano on earth—one which we may feel reasonably assured will soon awaken from its slumbers. Nor is this sensation at all diminished as we put our hands within the numberless little cracks with which the surface is riven and feel the heat ascending. What seemed to be a comparatively smooth floor from above turns out to be exceedingly rugged and by no means easy to travel over. It is a wilderness of writhing and seething currents and eddies of basaltic obsidian and broken folds of pahoehoe which shiver and crack beneath the feet as we wander on. It is all as fresh as if it had solidified only yesterday. It reproduces with exactitude the appearance of the lava upon the borders of the new lake in Kilauea. Here and there a whiff of sulphurous fumes reminds us of what is beneath our feet. But for these indications we should hardly be conscious that the ground beneath is less safe than the granites or sandstones of any other region.

We may ascend the black ledge upon the southern side in the vicinity of the little cinder cone. Here the rocks appear to be somewhat older, although many patches of extremely fresh basalt are seen on every hand, welling up from numberless fissures. Near the base of the great wall of the amphitheater we come upon numerous wide cracks, too wide in fact to be crossed, and which it is necessary to go around. Right at the base of the southern wall are several long, narrow, sunken blocks, which are lower than the level of the black ledge and which have evidently dropped in. This phenomenon is repeated at many points around the base of the main precipice, indicating that the dropping action repeats itself frequently. The cliffs at many points are so nearly vertical and so rifted with vertical cracks that they seem liable to fall down in terrible ruin at any moment. The base of the cliffs is littered with numberless masses which have fallen in this way in times past. At the southern end of the amphitheater is another and smaller caldera, connected with the main one by a narrow passage frightfully incumbered with fallen rocks of great size, which have tumbled from the cliffs above. It stands to the principal caldera in precisely the same relation as the pit of Poli-o-keawe to Kilauea. And, in truth, it is remarkable to note how closely Mokauaweoweo and Kilauea resemble each other. As if to make the parallelism exact, a stream of lava issuing from the neck which separates the two, flows into the depths of the smaller one.

A journey throughout the expanse of the caldera, with a study of its floor, discloses more questions than answers. First and foremost is the question, how was this wonderful depression formed and how is it maintained? Fifty or sixty years ago, no doubt, most geologists would at once have conjectured that it was blown out by some stupendous explosion hurling the fragments far and wide over a vast area, and that the cavity was thereafter refilled by the rising lavas to a level of nearly a thousand feet below the shattered rim of the crater. Such an explanation would be wholly unsatisfactory to the geologist of the present day. Cataclysms are not postulated in modern geological science without evidence amounting to positive proof, and no evidence of any such convulsion exists here. All around the base of the encircling cliffs may be seen the cracks and planes of cleavage; and the bays and recesses in the scarped wall disclose similar large joints cleaving parallel to the front of the cliff, indicating the action of the shearing force set up by the undermining of the masses which we may infer to have formerly extended over the cavity. Huge wedges of older volcanic beds are sunken even below the general level at the foot of the cliffs. All these facts convey to the mind the idea that the summit of the mountain has sunken in. It is an easy inference that the lavas beneath have remelted the underlying portions, and that the reliquified basalt has been tapped and drawn off by eruptions breaking forth far down the slopes of the mountain. This explanation, however, does not cover the entire ground of inquiry. Has this great pit been formed a few hundreds or a few thousands of years ago, or has there always been a similar pit in this locality maintained by the same causes throughout the whole of the immense period covered by the growth and development of Mauna Loa? That it is by no means an isolated or unique occurrence is proven at once by the exact homology of Kilauea and of Haleakala, and homologies less exact and more or less modified may be found in other great volcanoes. On the whole, it seems to me more probable that such a formation may belong to the middle and later stages of the growth of the volcano rather than to the earlier ones. For it is apparently associated with the habit manifested both by Kilauea and Mauna Loa, as well as other volcanoes, of discharging their lavas at levels much below the summit of the main hydrostatic column in the central orifice. This is much more likely to occur in a very large and lofty volcano than in a small and low one. All of these great pits give indications of progressive enlargement by the successive cleavage and dropping in of large spalls faulted from the faces of the encircling cliffs. The idea of a stupendous explosion blowing off the top of the mountain is wholly at variance with the peculiar habits of these volcanoes, for of all volcanoes they are the least demonstrative and violent in their respective activities. On the whole, it seems more probable that these pits are very ancient, and have existed certainly throughout all of the later stages and perhaps through the middle stages of the development of the several volcanic piles.

Another very interesting question presents itself here. While the fires are active a large lava lake is boiling and surging within the black ledge. If in reality the fires ceased to burn because the lavas were tapped and drawn off at a lower level, why should not the liquid have all receded downwards in the pipe and have left a large empty chimney hole of unfathomable depth? As a matter of fact, the frozen surface of this lava lake forms a nearly continuous plain across the whole lower floor of the caldera. It seems to have frozen over without any change or sinkage in the height of the summit of the lava column. On the one hand, the activity of the lake which precedes and the quietude which follows an eruption upon the flanks are too well attested to leave any doubt as to a sympathy between action in the lake and the lateral eruption. On the other hand, laws of hydrostatics seem to be so simple and inexorable that we cannot fly in the face of them. The action of Kilauea in these respects is much more consistent. It has been repeatedly observed that after the eruption upon the flanks of Kilauea the lava columns within the lakes disappeared for a time, but soon after gradually came up again. It is possible that the same thing may have occurred at Mokuaweoweo. It should be borne in mind that Kilauea is under almost constant observation, while Mokuaweoweo is rarely visited. If the behavior of the latter vent is the same, the difficulty will disappear.

There is still another enigma. Eruptions from the summit platform above are very frequent. Three or four of these have happened within the last thirty years, and although they might be reckoned large eruptions at Vesuvius or even at Etna, they are small in comparison with the enormous outbreaks which have flowed for many months and which have been the objects of general interest and attention. Since no systematic observations have been made upon the eruptions of Mauna Loa, and as these smaller ones seldom reach even as far down as the upper limit of vegetation, they have not attracted much notice, and many of them may have escaped observation altogether. This is all the more likely because Mauna Loa is usually enveloped in clouds, especially during the winter season, and the summit is rarely visible from below. In truth, the upper dome of Mauna Loa, as might be expected, consists of thousands of these minor eruptions piled one above another, and wherever we ascend to the summit we observe on every hand innumerable *coulees* terminating at varying distances above the timber line. How is it possible for an eruption to take place upon the summit platform of the mountain while a great central vent or pipe is open to the air within the caldera nearly a thousand feet below? Perhaps some gleam of light may appear when we recur to the fact that lavas are effervescent compounds highly charged with elastic vapor, and whenever a fissure or rift is opened for their escape the frothing of the liquid by the expansion of its occluded gases may be sufficient to carry a portion to a higher level than the main orifice. This would imply that the rising subterranean lava columns have many passages which are neither in connection with each other nor with the main central pipe unless such connection exists at great depths. This explanation may be doubtful.

Two days are all too short a time in which to go over carefully the various points of interest disclosed within the caldera and upon the surrounding summit, but the hardiest traveler will shrink from passing a second night in that dreadful solitude. The wind blows with a keen, biting blast, from which there is no shelter. There is not an ounce of combustible matter within twelve miles, and the only food is that which we bring with us. The animals must cower and shiver, and howl piteously throughout the night, and the guides or packers, accustomed only to the tropical climate of the sea shore, are hardly fit to endure the hardship. But the sublimity of the desolation and the deep sense of solitude, half pleasing, half terrible, make the experience a fascinating one.

Upon our downward path the all-absorbing theme was the wonderful display of cloudwork spreading out beneath us. The upper part of Mauna Loa is far above the ordinary realms of cloudland. The constant trade-wind is full of clouds of a peculiar habit. Their lower surfaces vary in height from 2,000 to 4,000 feet, while their upper surfaces seldom reach above 8,000 feet. From the upper dome of the mountain we look down upon them from regions which, in the summer-time, are almost always clear. At sea-level the atmosphere is always hazy, whether over the ocean or over the land, so much so that it is rarely possible to see one island from another, even across a channel fifteen or twenty miles in width. But on the mountain tops the purity of the atmosphere is marvelous. From the summit of Mauna Loa the dome of Haleakala on Maui, a hundred miles away, is defined so sharply and clearly that every detail of sufficient size is seen with the utmost distinctness. It rises out of a turmoil of clouds like a flat mound or dome rising above fields of snow. The aspect of the clouds when seen from above is in the strongest contrast with our notions of them when viewed from beneath, and a more inviting playground for the imagination it is impossible to conceive.

I was much impressed with the fact that the trade-wind is not felt on any of the high mountains of these islands at altitudes above 7,000 to 8,000 feet. The upper part of this mountain is in a region of complete calm excepting the uppermost 2,000 feet, where a gentle wind usually blows in a direction opposite to that of the trade-wind. Storms and gales of great power sometimes prevail upon the summit, but there are no observations which may enable us to estimate their frequency. The accounts given of them by those who have experienced them indicate that they are quite independent of the trade-winds. Most of them seem to come from the southwest and northwest. It is certainly remarkable that a wind so powerful as the trade, and covering so wide a zone, should be confined within such narrow vertical limits.

CHAPTER VI.

THROUGH PUNA TO HILO.

Four days were spent at Waiohinu in refitting for a second expedition and in developing the dry-plate negatives exposed at many places along our journey. The negatives, for the most part, proved to be failures. The atmosphere of the islands is very obnoxious to out-of-doors photography, especially with the dry plate. The perpetual haze which is in the air renders it impossible to obtain a picture of details at a distance much exceeding one mile. This effect is much more pronounced upon the extremely sensitive dry plate than upon the wet, and proved throughout an insuperable obstacle to successful photography. The case, however, was reversed at high altitudes. There the atmosphere is extremely clear, and all the pictures taken at high altitudes proved to be fairly successful.

I contemplated for a second journey a much more extended line of travel than for the first, and determined to make the circuit of the entire island, and also to visit Mauna Kea. With a heavily-loaded pack-train I left Waiohinu and reached Kilauea easily in two days. The burning lakes and the floor of the caldera were revisited, and the photographs were again tried, with somewhat improved results, though by no means wholly satisfactory, and many surrounding points of interest were visited. Leaving Kilauea I took the trail leading into the district of Puna, which is the southeasternmost portion of the island. The road leads through a forest of ohia, with a heavy undergrowth of large ferns and shrubbery, and over fields of pahoehoe only partially covered with a scanty soil. The rainfall is very considerable, but so recent are the lava flows which have built up this portion of the island that the rocks are not as yet sufficiently broken down to form any great amount of soil. Two large cinder cones are noteworthy objects by the way, being situated along what appears to be a line of frequent rupture extending from Kilauea eastward. From the summit of the loftier one, which rises about 680 feet above the plain, an extended view of the country round about may be obtained. A magnificent view of Mauna Loa is disclosed and a portion of the caldera of Kilauea lying to the westward. Mauna Kea would have been in full sight had not the clouds intervened. To the eastward the gentle slope of Kilauea declines away to the sea, and a row of cinder cones is seen in the distance ranging along the same line of eruption. No less than seven large pits may be detected similar in character to Poli-o-keawe, some of which are even deeper, and one or two are quite as large, if not larger.

For eight or nine miles the trail steadily descends, but so gradually that it is only just noticeable. At length we reach the verge of a long steep hill, down which the trail zigzags among the rocky fragments to a platform 700 or 800 feet below. A few miles further on we are clear of the ohia forest, and find ourselves among the beautiful kukui or candle-nut trees with their bright green foliage and dense shade. Again the trail descends obliquely a long, steep hillside, which sweeps downward quickly to a broad, smooth platform near the level of the sea, which is now only two or three miles distant. On every hand are fields of pahoehoe half covered, or even less than half covered, with a hardy prickly grass growing in the merest film of soil. Where the road strikes the lowest plain the climate is dry and supports little vegetation. But by one of those rapid changes of climate that we encounter as we move from place to place on this island we may behold in front of us, and only a few miles away, one of those beautiful transformation scenes with which this changeful climate is associated. Trees of many varieties with strongly contrasted habits and foliage, blended in ever-varying tints, betoken the splendor and langnor of tropical vegetation. The sea-coast is margined in many places with abundant groves of cocoanut palms and dense thickets of pandanus or screw-pine. Just at sunset we make our camp by the side of a pool of slightly brackish water in a cocoanut grove. A little way off is a cluster of grass-houses, built in true native fashion except for the glazed windows, while among them is a white-painted board cottage and a little church, which also serves the purpose of a school-house. It is hard to say whether these structures built in civilized fashion improved the prospect or not. They certainly seemed out of place in a region where everything else had the aspect of tropical barbarism. They served, however, to remind us that we were in a region where all that is horrible and hateful in barbarism has been supplanted by much that is good in civilization, by the reign of civil law, the security of life and property, and the establishment of peace.

There is no portion of these islands where so much of the primitive character of the Hawaiians is retained by the people as in Puna. The district is seldom visited by white people, and I am informed that only two families of whites reside there. The native population is somewhat scanty and has undergone a great decrease within the present century, as in all other parts of the island. This decrease, however, seems to be due more to the emigration of the inhabitants to the large towns, like Honolulu and Hilo, than to the ravages of those diseases which are supposed to be the prime cause of the decay of the Hawaiian race. Many of the natives also go to other parts of the island, where they obtain employment upon the plantations and in other occupations. But those who remain retain considerable of their primitive character, spending the day in lounging, fishing, and visiting, living in grass-houses and subsisting principally upon fish and poi. On the other hand, they are amiable, hospitable, and peaceful to the last degree. They have civilized clothing, but often, as a matter of preference, go about wearing a shirt and a malo. Probably in no part of the islands have the teachings of the missionary produced a deeper and more lasting impression. Their village has a church for its most conspicuous structure, and on Sunday all the natives go to church with *furore*. So intense is the sabbatarianism that I found considerable difficulty in avoiding arrest and prosecution for riding through one of their villages on Sunday with a pack-train.

I was much pleased at the comparative neatness and order of the grass houses in which most of the natives still live. The furniture is simple in the extreme. The floor is covered with mats woven of lauhala (pandanus) leaves, and are scrupulously neat. Tables and chairs are seldom used except as luxuries. Food is eaten *à la Turque*, the family sitting cross-legged around the dish of poi. Most households possess crockery, knives, forks, and spoons, but calabashes made from large gourds are still used, and "fingers were invented before forks." I spent an hour watching an old kanaka making a calabash, with as much delight as when, an urchin of seven, I used to watch the cobbler mending a shoe or the wandering tinker grinding knives and scissors. Not a little suggestive were long rows of letters in their envelopes, stuck cornerwise into the slats to which the bunches of grass are tied to form the wall of the house. All natives of suitable age can read and write their own language, for education is compulsory. They correspond most vigorously, and the mail facilities are remarkably good considering the scanty population and resources of the kingdom. Every week the post-boy rides through from Hilo to Kau, via Puna and Kilauea, and back again. The saddle-bags are full of letters and weekly newspapers from Honolulu, printed in the Hawaiian tongue. This does not sound very barbaric, and in truth the Hawaiian is in all essentials as well civilized as the poor people of England or America. He owns his property in fee; he makes laws, executes and obeys them; he reads and writes; he has but one wife; he tills the soil and tends flocks; sometimes he accumulates wealth and sometimes he does not; he makes his will in due form, dies, and receives a Christian burial. In no land in the world is property more secure; indeed, I have yet to learn of any other where it is equally secure from burglary, rapine, and thieving, or those subtler devices by which the cunning and artful succeed in getting possession of the property of the less astute without giving an equivalent. All this is seen in Puna, which is no doubt the most primitive district in all the islands. The few relics of barbarism remaining are of the most harmless description and probably as good for the Hawaiian as any civilized customs he might adopt in place of them, and certainly not inconsistent with all the comforts and blessings of good laws cheerfully obeyed and well administered.

The Hawaiian race must ever be regarded as a very extraordinary instance of a primitive people taking on full civilization in the course of only two generations at most, and we might almost say in a single

GRASS HOUSE.

FOREST SCENERY—PUNA.

generation. It is seen most distinctly in the establishment and highly successful operation of a judicial system and statutory code founded upon those of New England and New York. In matters of general legislation the code is essentially a copy of the codes of those States. The judicial administration is on the whole excellent. The chief difficulty is in obtaining justices of the lower courts of primary and petty jurisdiction who possess the requisite experience and knowledge. These are for the most part natives, and necessarily so. Their intentions are always of the best, but their tendency is to construe law in accordance with their own notions of abstract justice in each particular case, rather than upon legal principles, and few of them are capable as yet of understanding the value and significance of precedents. But the higher courts are always open to appeal, and these are held by judges of American or English birth and education, and well qualified to dispense sound law and equity.

The respect of the native for statute law is very great, and the sheriff or policeman or assessor has no more difficulty in executing his process than in Massachusetts; if anything, he has less. The native of Puna enjoys the security of his property, his life, rights, and liberties by the same title as the American citizen and in equal measure.

Seventy years ago a state of social organization prevailed here which had its nearest counterpart in the feudal system of Germany in the ninth or tenth century, and a state of the arts equivalent to a low order of barbarism, though considerably above mere savagery.

So fascinating was the spot where my camp was pitched and its environment that I spent a day in rambling over the adjoining country, rather from a feeling of idle curiosity than from a desire to investigate any special object. Half a mile away through the cocoanut grove was the sea, where the long heavy swell of the ocean, driven for thousands of miles before the trade-wind, rolled grandly in upon the battlements of black basalt. Perhaps the strongest attraction was the pandanus groves. These trees have very long slender leaves, 3 or 4 feet in length, which grow in a spire around the trunk and branches, and curve over until they hang pendulously, giving the tree a habit not wholly unlike a weeping willow, while the foliage is very much denser. They cluster together very thickly, and form the most impenetrable shade of any tree with which I am acquainted. They have no need of soil, for they send down their roots among the moist rocks and clinkers, where no trace of soil is visible. They seem to me to have a peculiarly barbarous habit, well fitted for the horrible rites of savages. To make the illusion more complete I found within an opening in one of these large groves an ancient heiau or temple in a good state of preservation. A long slab, on which human victims were immolated, was pointed out to me, with a look of commingled awe and amusement, by a native dressed in a costume of civilization so complete and recent that I would not have disdained to wear it even here in Washington.

Hardly less attractive are the cocoanut groves. Indeed, Puna is certainly the home of the cocoanut. Nowhere else upon the islands does it seem to flourish so exuberantly. The tree also has, to my mind, a barbaric look and an aspect which is full of character. Its stem runs up to an altitude often as great as 60 or 80 feet, ending in a tuft of large fronds, 10 or 12 feet long, among which the cocoanuts grow in clusters, as many as 50 or 60 being found upon a single tree. These palms grow usually in abundance upon the sea-coast, though they are occasionally found far inland at heights of a thousand feet or more. But they are never abundant far away from the shore.

The inhabited portion of Puna is situated upon a flat plain of lavas with here and there a little soil and having but a very slight elevation above the sea. Two or three miles away from the coast the surface suddenly rises so abruptly that in many places the ascent is precipitous, and in the western part of the district the ascent is always very steep. Upon the summit of the escarpment or hillside runs that long line or axis of eruption which has already been spoken of, and from which in times past very many flows of lava have been outpoured and descended to the plain below. A few streams may still be seen which appear to indicate great recency in their periods of flow. But no definite epoch can be assigned to any of them, though the traditions of the natives declare that no king ever reigned in Puna without seeing large parts of his dominion overflowed. Throughout the entire district all of the lava shows more or less vegetation, but the climate is very moist, and it is well known that vegetation establishes itself quickly upon the lava beds where the rainfall is abundant and by no means waits for the formation of soil. As we move along the road to the eastern end of the island we find nothing but an almost unbroken expanse of rugged lavas mostly in the form of aa, or clinkers, and clothed with a most exuberant vegetation. So rough are these lava fields that without a road they would be impassable for animals and extremely difficult to cross on foot. Add to the roughness of the lava the jungle of vegetation which grows upon it, and the difficulty is even more than doubled. The main road through Puna leads along the sea-coast upon the southwestern border of the island. On the route comparatively little material is presented for geological study which may call for special comment. The entire distance is covered with enormous flows of lava, mostly in the form of aa, descending from the principal axis of eruption already spoken of and situated upon the high land to the northward. The character of this lava varies but little, all of it being basalt of the types normal to Kilauea.

Along the road there are about a dozen native villages situated in openings of the tropical forest, consisting mostly of grass houses, with here and there a somewhat more pretentious structure built of California lumber. As we proceed eastward the forest increases in density and also in variety, becoming at last extremely luxuriant and beautiful. One of the most attractive spots is an immense grove of cocoanut trees situated 3 or 4 miles from the easternmost cape of the island. Many thousands of these trees send up their snaky trunks to the height of 60 or 70 feet, and stand sufficiently close together to form a canopy of shade and to produce the effect of a forest. The ground is thickly cumbered with huge cast-off fronds and with myriads of nuts which are never harvested. In many places the pandanus groves are on either hand with their long weeping leaves and dense shade, forming an impenetrable jungle with their strange aerial roots.

As we approach the eastern cape of the island the road turns sharply to the northward, cutting through clinker fields densely clad with tropical forest. At last it emerges upon an open space where the rocks are completely hidden by a thick soil. On either hand may be seen several cinder cones, and among them a cluster of three which are almost merged together. At the base of this cluster is situated a fine large ranch indicative of thrift and comfort, where we found a most hospitable reception for the night. These cones are situated upon the principal axis of eruption, and from their summits we may see to the westward a line of old craters reaching out in the direction of Kilauea. A long period has evidently elapsed since these cones were formed, for they are considerably degraded by erosion and thickly clothed with vegetation to their summits. Within the crater of one of these cones is situated a beautiful pool of water about 170 feet in width, and the rankness of the vegetation about it renders the spot as picturesque as can well be imagined.

Five miles further on we come suddenly upon the great lava flow of 1840, at the point where it enters the sea. The width here is a little less than a mile, but farther inland we can see it spreading out to a width which must be nearly 3 miles. The sources of this eruption were about 12 miles distant, and have been described by Wilkes. The lava is highly olivinitic, and in this respect is rather exceptional among the lavas emanating from Kilauea. It very closely resembles, and indeed is undistinguishable from, the great floods which have come from Mauna Loa. At the point where the *coulée* reaches the sea three considerable cinder cones were thrown up in the same manner as those which are seen at the termination of the flow of 1868. Undoubtedly they were formed by contact of the molten lava with the water, and they are not distinguishable in character from the normal cinder cones which may be found encircling an ordinary volcanic vent. The lapilli of which these hills are built is all very fine, but perhaps not more so than that which occurs in many cinder cones which have been formed in the ordinary way. The sea has in great part demolished one of these cones and has made considerable ravages in the others. In a few years, doubtless, they will all disappear.

About 17 miles of easy travel brings us to Hilo. Little is to be seen along the route, except the luxury of the tropical forest, the beauty of which increases steadily as we approach the town. It is doubtful if its luxuriance can be surpassed by that of any other country in the world.

CHAPTER VII.

FROM HILO TO MAUNA KEA.

From Hilo I decided to make an advance at once upon Manna Kea and to visit the intervale between that mountain and Mauna Loa. Mauna Kea may be approached from many directions, the easiest lines of access being from the northwest and north. The approach from Hilo is the most difficult of all, because it involves the necessity of traversing the belt of forest which lies between the middle slopes of the mountain and the sea. No one can imagine the density and exuberance of tropical vegetation until he has seen it. In truth, the forest can be penetrated only by hewing a way through it or by traversing a route which has already been cut by main force.

It is well to point out here that the forest region of this island is regulated by the precipitation. The windward side has a very heavy rainfall, and a portion also of the western side is similarly favored. Most of the region under the lee of the island is arid, and in many places extremely so. Although vegetation upon the windward side is very abundant, even down to the margin of the sea, it never has that close impenetrable character near the sea-coast which it assumes further inland. The reason for this is not difficult to discern. The windward coast of the island is for the most part very abrupt, and the water which falls upon it rapidly drains away. The trade-wind striking the shore is deflected upward by the gradual ascent of the land, and at heights varying from 1,000 to 4,000 or 5,000 feet the clouds envelop the land in fog and yield an almost constant rain. The effect of this upward deflection producing a condensation of moisture is not so fully felt at altitudes below a thousand feet, and thus we have near the seacoast a margin of land which enjoys a great deal of sunshine, and even long periods of drought sometimes occur along the immediate neighborhood of the coast, while a mile or two inland it rains almost incessantly. The forest has its maximum density in the region of clouds.

The rainfall upon the windward side of Hawaii is phenomenally great. The mean annual precipitation as shown by the records extending through eighteen or twenty years ranges from 150 to 240 inches. This, however, is the result of measurements made near the sea-coast. Further inland it must be still greater, and may even attain more than 300 inches. Hardly a day passes at Hilo without a copious shower, and in the winter time long continuous rains always occur.

There are two routes leading from Hilo to Mauna Kea. One extends along the coast northwestward for about 30 miles, then turns abruptly upwards, striking the northeastern flank of the mountain. The other leads directly inland, and passing through the forest belt reaches the southern base of the mountain and the intervale between it and Mauna Loa. Each route has difficulties peculiar to itself. The first one leading along the coast strikes into a country which is deeply scored with very abrupt ravines and ridges. Here the land terminates in a cliff from 300 to 500 feet in height, plunging down into deep water; and against the base the heavy swell of the Pacific, driven before the trade wind, is constantly breaking. Along the front of this cliff near the water's edge no pathway is possible. The country can be traversed only by going up and down the walls of the ravines which at frequent intervals score the platform above. The sides of these ravines are very steep, and in many places have all the abruptness of cañons. With much labor, very fair trails have been cut zigzag in the sides, and sure-footed animals may go up and down with perfect safety, but with great labor. Within a distance of less than 30 miles there are upwards of 60 of these ravines of varying depths, and steadily increasing in dimensions as we go northward. The two last ravines into which the trail has been built are very impressive and picturesque. One of these, known as the Waipio gorge, has a northern wall about 1,400 feet high, the slopes probably exceeding 40 degrees. The beauty of the scenery consists more in the richness and luxuriance of tropical vegetation than in anything else, although the boldness and magnitude of the rocky walls are important elements in the picture. Many of these gorges carry living streams which are subject to frequent floods and which inundate very rapidly after the prodigious bursts of tropical rain.

In going from Hilo to Mauna Kea I declined the coast route across the gorges, and chose the much more direct line of approach passing through the forest. For two or three miles from Hilo the trail, if such it may be called, for scarcely any trail was visible, led through a country which was quite open and densely clothed with high grass. This grass is worthy of some little mention, for it is an exotic plant. Several accounts are given of the manner in which it was imported. Some describe it as a native of Holland, others as a native of Italy, and still others as coming from the Cape of Good Hope. It is said to have been brought to the island by accident; that the dried grass containing the seeds was used as the wrapping of bottles containing wine or oil; that the seed accidentally scattering at once took root, and finding the soil and climate specially adapted to its growth, spread with marvelous rapidity, and flourished with such vigor that in the moist districts of the island it has almost exterminated all other grasses. In its green state it is hardly fit for pasture. The cattle and horses eat it, but apparently get very little nourishment from it; for leaner and more cadaverous-looking horses and horned cattle it would be difficult to find than those which are pastured in the vicinity of Hilo. So dense and high is this grass that a passage through it on horseback is attended with extreme labor. It looks very green and inviting, but its very inferior character as a food for animals is abundantly demonstrated. It is said, however, to be very much better in the form of hay than when green. A dry climate is not well suited to it, and in such localities other grasses appear to hold their own. Perhaps the best variety is one which was brought from Mexico early in the century, about the time that horses were first imported. It is called, locally, *maniania* grass, and wherever it grows forms the richest and most velvety sward imaginable. It is highly nutritious and animals are very fond of it. It flourishes best in a medium or very slightly arid climate. It was once universal all over the island, but the Hilo grass in all the wet districts of Hawaii has completely exterminated it.

Upon the outskirts of the village of Hilo we find the end of the great lava-stream which flowed the year before my visit. It is typical pahoehoe. From a convenient standpoint in the vicinity we can see the last 3 or 4 miles of this stream, spreading out with a width of nearly a mile over the broad, open, grassy plain which lies just west of Hilo. The view of it is at length lost where it emerges from the forest. So flat is the country just here that by a common optical delusion the lava seems to have flowed up hill, though in reality the descent from the forest to the end of the stream may be anywhere between one and two feet per hundred. The slope, however, is exceedingly small. Within a half mile of the termination the thickness of the lava sheet appears to be very small, not exceeding, I imagine, 20 feet, and generally less. The numberless mounds or bosses of pahoehoe were all formed in detail in the manner already described, by repeated outshoots of streamlets from underneath the hardened crust behind. As these belches of lava cool they exclude the occluded steam, and the mass swells up by the formation of myriads of vescicles, and often also by the formation of great hollow blisters underneath. The supply of fresh lava during the last part of the eruption seems to have been quite copious, for the advance of the stream was nearly 300 yards per day.

The people of Hilo had concluded that there was no hope for the preservation of their beautiful village. The advance of the lava straight towards the town had been uniform for several months, and it was possible even to compute the number of days which would be required at this constant rate of progress to accomplish the destruction. As it drew near all portable property was packed up for removal, and many people would have sold valuable realty for a few dollars if purchasers could have been found. At length the end of the stream approached within about two days' march of the upper street. Already two long arms had begun to reach out divergently from the end of the flow, one extending as if to reach around the southern part, the other as if to reach around the northern part of the town, and finally to clasp the whole in its fiery embrace. Suddenly, without premonition, the movement ceased and was not renewed.

This eruption began, as before remarked, in November, 1880, and lasted until October, 1881. The eruptions of 1852 and 1855 broke out near the same point on the upper dome of Mauna Loa as that of 1880, and pursued closely adjoining and parallel courses. That of 1855 was much larger and that of 1882 a little smaller than this one. The length of the last flow (1880–'81) was nearly 50 miles, but its course is somewhat tortuous.

Three miles of travel through tall Hilo grass growing in a muddy soil brings us to the verge of the forest. Years ago a trail leading from Hilo up into the central wilderness of the island was cut through the forest and corduroyed. The trees used for the corduroy were trunks of the great tree-ferns which form a large part of the undergrowth of the forest. These are soft, spongy, and perishable, and lasted but a very few years. They quickly became rotten, and wherever they were laid the trail has become worse than it would be if they had never been put there. The effects of the incessant rain are now abundantly visible, and that to our great discomfiture. The trail is a mixture of rocks, mire, and fragments of rotten fern-trees. Progress is difficult and extremely harassing. Every few rods some poor animal sinks his fore legs or hind legs into tough, pasty mud, and must be unloaded and pried out. Four miles of this kind of travel was accomplished in the space of about six hours. Suddenly and without warning a sharp turn of the trail brought us upon a wide expanse of naked pahoehoe. The relief was indescribable. Nobody would pretend that pahoehoe is pleasant traveling. It is good only in comparison with clinker-fields and forests. The exchange is that of misery which is intolerable for misery which can be borne readily by the exercise of patience. The animals being exhausted by the desperate struggle, we at once made camp upon the lava rock, finding a pool of swampy water hard by.

We had landed upon the termination of the great flow of 1855, the grandest of all the historic eruptions of Mauna Loa. The next day we had an opportunity to observe and appreciate its immensity. Our route lay upon the upward course of this flow, which soon widened out on either hand until the forest was miles away from us in both directions. Already a few straggling ferns and other humble plants have begun to take root upon its surface, but without a vestige of soil. Except for these stragglers all is now bare rock, rolling in heaps and mounds, twisted ropes and huge wrinkles, with now and then a network of cracks rifting the mass into fragments, and large holes where the arch over some great lava pipe has fallen in. One characteristic of this great flow is the exceptional unevenness of it and the large size of the mounds and hills formed by the pahoehoe. It seems to lie very much thicker than in most other eruptions. In many places it has formed high hills or ridges, and everywhere there are abundant indications that sheet after sheet of lava was piled up to form its final mass. The width of it a few miles above its extremity could only be estimated roughly by the eye, and seemed in many places to exceed six miles. In the course of an hour the forest was dim in the distance on either hand, the tall ohia trees appearing like mere shrubs.

As I looked over this expanse of lava I was forcibly reminded of some of the great volcanic fields of the western portion of the United States, where the eruptions are of such colossal proportions that they have received the name of massive eruptions. Richthofen, after studying many of these lava fields in California and Nevada, was led to the conclusion that they had burst forth from great fissures, inundating large areas of country with fiery seas of basalt. He was led to contrast the immense volume of these rocks with the comparatively insignificant streams which have emanated from Vesuvius, Ætna, and other modern volcanoes, and concluded that the incomparably grander overflows of Western America must have occurred under circumstances differing widely from those of ordinary volcanic eruptions. Although the volcanic rocks of Western America may be considered as very well exposed as compared with rocks of equal antiquity in other portions of the world, they would be regarded as relatively obscure by any one who has had an opportunity to inspect carefully the recent lavas of Mauna Loa. I am by no means certain that Richthofen's conclusions are wrong. But here is a lava flow, the dimensions of which fully rival some of the grand pliocene outbreaks of the West, which demonstrably differs in no material respect, excepting in grandeur, from the much smaller eruptions of normal volcanoes. The flow lasted for thirteen months without interruption, and in that period it is easy to see that an enormous volume of fluent lava could be disgorged from an orifice of no very extravagant proportions. In estimating the volume of materials composing this flow there is one unknown factor, namely the thickness. Probably this can never be ascertained with a satisfactory approach to accuracy. It is extremely variable, and the configuration of the country which it deluged is wholly unknown in detail. The surface of the flow has not as yet been accurately surveyed, and its horizontal dimensions have been subjected only to eye estimates, which are extremely untrustworthy. The want of proper data, therefore, makes it unwise to venture an estimate of its mass. Some impression, however, of its grandeur may be derived from the statement that for a distance of 20 miles from its termination the average width of the flow cannot be less than four or four and a half miles. The axis of the main stream from its source to its termination is a little more than 45 miles in length. The thickness of the stream in many places is very great, probably exceeding 250 feet, while the average may not exceed 100. Its final solidification has left the general surface extremely irregular, being piled up frequently in ridges or hillocks 50 feet high or more. By far the greater part of this mass is pahoehoe, and it was formed no doubt in detail after the manner which has already been described.

A little more than 20 miles from the end of the flow we found ourselves confronted by a high barrier of clinkers stretching far out towards

END OF THE LAVA FLOW OF 1881.

MAUNA LOA FROM THE NORTH.

the base of Mauna Loa on the left and plunging into the forest on the right. Turning sharply to the right the trail crosses several spurs of this ridge of clinkers and at length leaves the lava field and enters the forest. The character of the forest is now greatly changed. It is no longer a swamp and jungle. We have gained an altitude of about 5,500 feet, and although we are not wholly above the wet region we are in one which is considerably dryer than that which is occupied by the main forest belt. The soil in the summertime is generally dry, and the undergrowth is so moderate that it offers little obstruction to progress Winding through the forest we come frequently upon open parks densely clothed with mountain grass. The trail ascends slowly but steadily, and as we progress the trees become fewer and the parks larger and more numerous. Numberless trails of wild or half wild cattle traverse the country in every direction. The soil is abundant, but so too are the ledges of lava and fragments of clinker which project through it. Ascending a rocky shelf, Mauna Kea discloses its magnificent mass in close proximity on the one hand, while Mauna Loa, more distant and yet more grand, rises sublimely upon the other. The prospect towards Mauna Loa is desolate in the extreme. The wide intervale between the two mountains is an enormous expanse of ominous black lava, mostly *aa* and clinkers which seem to bid defiance to all access. The sides of the mountain are everywhere streaked with descending tortuous bands indicating the positions of more recent lava flows. Where these strike the plain below they spread out into wide fields of clinkers. The fact is a significant one, and the explanation does not seem difficult. Upon the mountain slopes the lava runs with great, velocity, and the streams are correspondingly narrow. But when it strikes the nearly horizontal plain below its velocity is checked and the liquid accumulates in great volume, becoming viscous by cooling. Its flow is greatly retarded and yet the mass is sufficient to enable it to move with a slow motion analogous to that of a glacier. When the viscosity of the lava becomes very great it is in a condition which enables it to yield to strains of a certain amount, but if that strain is exceeded it is crushed and ground up. The movement which takes place at this stage is partially a plastic yielding, more particularly of the interior and hotter parts, and partly a shattering and grinding up of the outer stiffer and colder parts. This glacier-like motion, however, is possible only with very large masses of the lava which still retain a sufficient quantity of heat to maintain a plastic condition. Persons who have witnessed the movement of a clinker field in the last stages of an eruption describe it as being so slow as to be quite imperceptible until it has been watched for a long time, and as being attended with a cracking noise which comes in vollies like the report of musketry.

Turning around with Mauna Loa at our backs, the majestic pile of Mauna Kea rises immediately before us. The contrast is very great. The eye is instantly caught by the large number of cinder cones which everywhere stud its surface, from the summit where they cluster thickly, down its flanks to the plain below. All of them are symmetrical and normal in their outline, and in an admirable state of preservation. They are truncated at their tops, showing the existence of regular craters within the truncated portions. Some of these cones, by a careful eye estimate and comparison with known magnitudes, appear to be more than 1,000 feet in height and more than three-fourths of a mile in diameter. The number is too great to be easily counted. They are most numerous upon the summit platform, but they are very abundant, not only upon the immediate base of the mountain, but at all intermediate zones, and they ramble away far beyond the base like a crowd dispersing from a common center. The general form of the whole pile of Mauna Kea is notably different from that of Mauna Loa. Its slopes are much greater. And yet they are very far from being so abrupt as those which are found in the majority of the grander volcanoes of the world. Nowhere do they appear to exceed fifteen to eighteen degrees, except upon the flanks of the cinder cones, and the average profile upon the side here in view is about twelve degrees. The northern front of the mountain, which is not visible, has a slope considerably greater. Comparing this with Mauna Loa, we find that the average slope on the steepest flank of the latter mountain nowhere exceeds seven degrees, and in the longer ones it is only four degrees. Yet, in comparison with other great volcanoes, Mauna Kea is rather flat and obtuse.

Its composite character is disclosed at once. It has no dominant central peak or cone like Etna, Shasta, and Teneriffe, which completely overpowers all other features, but it has been accumulated by eruptions from numberless vents, which are spread out over a very large area and cluster most thickly at the central and highest part. Upon the summit are many large cinder cones grouped closely together and planted upon a well marked summit platform. But it is impossible to select any one out of a dozen of these cones which can be confidently pronounced largest, nor is it possible to say which out of half a dozen is the highest. Cones even larger than those upon the summit are seen at varying altitudes upon the flanks.

Glancing back once more at Mauna Loa, not a single cinder cone of normal type is anywhere visible upon all its mighty expanse. Far up towards the summit are two or three minute pimples, which, upon examination with a strong field-glass, convince us that they were originally intended for cinder cones, but that the attempt was abandoned in a preliminary stage of the experiment. All of that stupendous pile, so far as visible, is built of streams of flowing lava. But Mauna Kea consists largely of fragmental material. What proportion of its mass is thus composed of fragmental matter can only be guessed. But the percentage is no doubt great.

The lavas of Mauna Kea will be alluded to more in detail hereafter. At present it may be remarked that nowhere in this part of the mountain are its lavas well exposed. The volcano has been extinct for many centuries, and although the degradation on this side of the mountain has made comparatively little progress, we shall soon find reason for believing that the epoch of final cessation, historically speaking, is quite ancient. The impression produced is that the period which has elapsed since the last sign of activity should be reckoned by thousands of years rather than by hundreds. Soil is everywhere abundant, and no fresh-looking rocks are known. The dense forest comes up only to the level where the steeper part of the mountain begins its ascent; that is, to altitudes varying from 5,000 to 6,000 feet. Above that are many scattering groves with a gradually increasing proportion of open spaces. Up to an altitude of nearly 10,000 feet the mountain is clothed with long mountain-grass, which has a pale yellowish color. The cinder cones have that faint reddish cast often assumed by basaltic lapilli which has long been exposed to weathering.

Winding onward by a rough stony trail, where old rotten clinkers and slabs of weathered basalt project up out of the soil, we at length reach a pool of stagnant water, where we make camp. Just before reaching camp the way was somewhat obstructed by a thicket of thorny bushes which at once aroused the keenest interest. They were apparently raspberries, but *such* raspberries! The bushes were gigantic and the fruit equally so, the berries being over two inches in length and an inch in diameter. Conceive our ordinary pale red garden raspberries magnified two and half to three times in linear dimensions whether in stalk, leaf, or fruit, and we shall have a very good idea of its appearance. Its flavor, however, was somewhat inferior, though by no means unacceptable. The taste of the fruit is almost exactly the same as our common Lawton blackberry. The abundance of fruit was remarkable. For two or three miles the banks and hillsides were covered with them and they could have been gathered by thousands of bushels. They were growing at an altitude of about 6,000 feet, where snow frequently falls in winter and where the climate probably does not differ greatly from that of the coast range of California; though I presume this climate is rather the more equable of the two, being cooler in summer and perhaps a trifle milder in winter.

The journey from Hilo had been a very long and arduous one. Unpleasant as was the struggle with the forest, the journey of twenty miles over pahoehoe, so coarse and rough as that of the flow of 1855, proved in the end to be almost as harassing to the animals. The foothold upon the rocks is all that could be desired, but the constant ascent and descent of the smooth rounded hummocks produced an incessant lurching and strain upon the animals the effects of which were now manifest in the shape of sore and scalded backs. Two days' rest was deemed absolutely necessary to recuperate the sore, weary, and half-starved brutes. I occupied the time in tramping over the rolling hills and half-concealed lava beds around the base of Mauna Kea, and in exploring three or four long caverns or ancient lava pipes, which are quite as common here as they are upon Mauna Loa. No results of any importance attended the investigation. Many specimens of rock were picked up and examined superficially. They have no great variety, but at the first glance they show a well-marked difference as compared with those from Mauna Loa. Olivin is abundant, but is never seen in such excessive quantities. On the other hand, the feldspars are present in great quantity in well-marked tabular crystals, and many large crystals of augite occur. The groundmass in the majority of cases inclines to bluish gray instead of being greenish black, as in most of the lavas of Kilauea and Mauna Loa. In short, they are true basalts, approaching more nearly the normal type than those we have hitherto seen. The methods of flow are apparently quite similar to those seen on Mauna Loa. The two forms, pahoehoe and aa, are as distinctly represented and yet there is some difference, especially in the case of aa, but a difference which I should find it extremely difficult to define.

CHAPTER VIII.

MAUNA KEA.

After two days' rest and recuperation the ascent of Mauna Kea was determined upon. The summit is easily reached from the southern side, so easily in fact that no great precaution is necessary in the choice of routes. Still, some routes are much easier than others, and it was thought best, in view of the long and tedious character of the ascent, to take a guide familiar with the mountain. I found a native who had been to the summit many times and who had hunted sheep, cattle, and goats all over its southern flanks. At daylight the party was in motion with three pack animals carrying photographic apparatus, provisions, and also blankets, in case it should be found necessary to spend the night upon the mountain top. The guide went afoot from preference, a most unusual thing for a kanaka, while the rest of the party were well mounted.

Our camp was situated at an altitude of about 5,670 feet, and the top of the mountain was more than 8,000 feet above us. Two hours were spent winding deviously among the foothills and cinder cones around the base of the mountain before the principal slope of the mass was reached. The platform consists of lava beds in a somewhat advanced stage of decay and having much the same character originally as the lava fields which make up the gentle slopes descending away from Kilauea. There are the same alternations of pahoehoe and aa, but the roughness has been greatly mollified by weathering and by the formation of soil. In many places, especially at the foot of steep slopes, the soil has accumulated to a very considerable thickness, having been washed down from above, and lies in heavy banks. Erosion also has begun its work. Here and there we crossed sharply cut ravines of small depth scoured in the rocks by the torrents. As yet no perennial stream exists on this side of the mountain, but the evidences of frequent spasmodic floods of great power and volume are often encountered. As we reach the principal slope the ascent becomes very rapid, but by no means uniform. Here for a few hundred feet it rises so rapidly that the animals struggle and strain. There it is so gentle that we may jog along at a trot. With increasing altitude the slope becomes greater, and at last we dismount to ascend on foot a continuous slope at an angle of more than twenty degrees. We do not leave vegetation behind us until we have attained an altitude of nearly 11,000 feet.

Most of the route lies through an alternation of rugged fields of lava, which show less and less soil the higher we ascend, and the fine lapilli of the cinder cones, into which the feet sink deeply. The flanks of these cinder cones are never excessively steep, but owing to the very loose character of their component materials the ascent becomes toilsome and very protracted. The cones also become more abundant as we approach the summit. They show no signs of decay as yet, except, possibly, a little weathering of the lapilli in the upper layers, which have turned red and brown, while at some little depth the color is still black. It is worthy of note that the lapilli of basaltic cinder cones are sometimes red when first ejected, though more frequently they are black, the color depending, I presume, upon weathering. The iron constituents have the form of protoxide or peroxide. Weather usually converts the iron sometimes to peroxide, sometimes into the hydrated form. Many cinder cones, however, preserve, for an indefinite period, even until they are half obliterated, their original black color. In the cones of Mauna Kea the lapilli as originally ejected were, no doubt black, but have superficially changed to red or brown. All of it is comparatively fine and no large pellets are seen.

About one o'clock, after seven hours of travel without a halt, we reached what may be termed the summit platform, which has an altitude varying somewhat with its inequalities, but averaging probably 12,500 feet. This platform is about five miles in length and two miles in width, with a slightly pronounced ridge running along its axis. Upon this platform stand about a dozen large cinder-cones, from 700 to 1,000 feet in height, carrying the extreme apices of the mountain very nearly to 14,000 feet. It is difficult to judge which of these cinder cones stands highest. But it soon becomes apparent that this distinction belongs to one of a group which are clustered thickly together near the western end of the platform. Towards these we direct our steps.

The aspect of the lavas beneath our feet now becomes somewhat different from those seen lower down the mountain. They are lighter colored and some of them are much more compact. A fragment when struck rings like clinkstone, and on being broken shows a dark, but very compact fracture and an entire absence of the vesicles which are universal in the lavas which we have hitherto seen. Some are vesicular, others glassy or obsidian-like. It is interesting also to note the effect of weathering upon the summit. These lava beds have evidently lain for a long time exposed to the action of the elements. In a few places are to be seen traces of running water. But for the most part the weathering simply amounts to a slow decay and dissolution of the rock in place Some of the sheets have been broken up into small fragments, and by the gradual dissolution of the exterior portions the angles have become rounded and the fragments smoothed off. In one place we crossed what was once probably an old sheet of lava. This is now reduced to a mass of rounded stones separated by considerable intervals.

As we approach the western end of the platform we gain notably in altitude, and at length find ourselves in a spot where in almost every direction we are hemmed in by large cinder cones towering to a considerable height above us. Here we halted for a midday camp. We brought up a few sticks of wood to build a fire, and enjoyed a cup of coffee, a few slices of bacon and some bread. The guides suffered somewhat with mountain sickness, and the animals betrayed the effects of the unaccustomed altitude, for we were more than 13,000 feet above the sea. There is no difficulty in ascending the summit cones which are composed of fine loose lapilli and about 800 feet in height. The prospect was a total disappointment. The country below was completely buried in clouds, out of which the mountain rose like a great island. But to the southward was the mighty dome of Mauna Loa, rising above the clouds which floated about 6,000 feet below the summit. It seemed very near, though in reality it was about twenty miles distant. The great caldera was distinctly seen with portions of its encircling wall. There is a partial opening or gap in this caldera towards the north which enables the observer from Mauna Kea to look into it. And so clear is the atmosphere at these high altitudes that with a good field glass many details of the rock faces are easily discerned. To the southwestward and rising about 2,000 feet above the clouds was the summit of Hualalai, presenting an aspect quite similar to the summit of Mauna Kea, but upon a smaller scale. To the northwestward the dome of Haleakala, about eighty miles distant was in full view. By means of a field glass it was possible to discern easily the cliffs inclosing its vast caldera, and one or two of the cinder cones within it. A purer atmosphere than that which prevails here at high altitude, it is impossible to conceive. Even the summit of Haleakala is seen in its natural colors without any of the adventitious tints usually imparted to distant objects by a hazy atmosphere. Now and then a glimpse is caught of some small portion of the country below from momentary openings in the clouds. Upon the leeward side of the island short stretches of sea coast are here and there disclosed, but from so great an altitude they have a strange visionary aspect.

Several hours were spent in photographing and in rambling about the platform in search of whatever might be found. Hard by the noonday camp is a mass of very light-colored lava which seems at first to have a constitution notably different from the very black almost ultra basalts to which we have thus far been accustomed. It is exceedingly compact and fine grained and has a very light gray color. The fresh fracture, however, is notably darker than the smooth weathered surfaces. It has been called a feldspathic rock, meaning, I suppose, a rock more nearly allied to the trachytes than to the basalts. Other observers have called it phonolite, probably because it is highly resonant when struck. But the term phonolite is now used by lithologists to indicate a special and limited group of rocks having a tolerably definite chemical constitution and possessing nephelin as its most characteristic mineral. This light-colored rock of Mauna Kea, however, is undoubtedly a basalt possessing an abundance of triclinic feldspar in exceedingly minute crystals and without olivin. It appears to be identical with a very large proportion of the basalts occurring in the western portion of the United States. This rock was used by the primitive Hawaiians for making their stone implements, for which it is very well suited, being very hard, tough, fine grained, and free from vesicles; and it flakes readily. Hard by are abundant vestiges of the work of manufacturing weapons and tools; and incomplete products in all stages of manufacture, with large quantities of flakes, lie scattered about.

No signs of any recent volcanic activity are to be seen. All the lava beds look old and greatly weather-worn. In some of them the decay and disintegration are so far advanced that they are reduced to mere heaps of weather-beaten fragments. How these lava sheets have thus been torn to pieces, as it were, and reduced to piles of moldering ruins I can explain only by suggesting the action of frost and ice filling the cracks and wedging the pieces apart by expansion. To this, however, should be added the wasting away of the pieces by the solvent action of the rains. A few hundred yards from our noon camp is the head of a ravine which has been scored to a considerable depth by the unmistakable action of running water. Thus erosion has made a good beginning here, and under circumstances where its action is undoubtedly slow and spasmodic. This ravine has at one part a depth of nearly 70 feet, and is exceedingly rough and much obstructed by fallen fragments. The cinder cones, however, do not appear to have suffered much from the ravages of time. Their preservation is no doubt due to their open, porous character. The rain can never fall fast enough to start a torrent or even a minute rill upon their surfaces, but sinks into the interstices at once. Everything indicates that a long period has elapsed since these vents became silent.

The temperature at the summit in the daytime was rather mild, being about 50° F. The air was calm, only a very light breeze blowing. But we knew quite well that the temperature would fall greatly during the night time; and the lightly-clad kanaka is not fond of cold. As a minute exploration of the summit promised little of special interest beyond what had already been seen, I decided to seek a lower altitude to pass the night. As we started, the day was drawing towards its close, and as we reached the verge of the summit platform the sun was near the horizon. Meantime the clouds to the southward had dispersed, revealing the whole northern side of Mauna Loa, which rose in indescribable majesty before us. Through the clear, pure atmosphere every detail was visible. Innumerable recent lava streams could be seen stretching their tortuous courses from the upper dome down to the plain below, spreading out in enormous fields of blackness and roughness. Three long streaks in particular attracted the attention. One upon the northwestern side, starting from a point a little below the summit, reached down the slope into the broad intervale between Mauna Kea and Hualalai, and vanished away in the distance towards the sea-coast. This I had no doubt was the flow of 1859. Far to the left, upon the northeastern slope of the mountain, could be seen two streams which had flowed out from a year to a year and a half before. The one emanating from the point east of the mountain was the stream which first broke forth in November, 1880, and rushed rapidly down the slope directly towards Mauna Kea. The other, which was the last of three distinct streams from this eruption, started from a point lower down the mountain, flowing northeasterly then turning towards Hilo. Many other streams were distinctly visible, wearing an appearance of recency. Down the main slopes of the mountain these floods are comparatively narrow, having widths which might be from half a mile to a mile. But as they reached the plain between the two great volcanic piles they spread out into immense floods, which are mostly aa. The appearance of the plain thus deluged by the frequent outpours from Mauna Loa is black, desolate, and horrid in the extreme. They end very abruptly upon a sinuous line, where they meet the ascending slope of Mauna Kea.

The sun disappears and the brief twilight follows. At length we enter the clouds and move on in the mist and darkness, reaching camp a little before midnight.

In the afternoon of the day following the ascent of Mauna Kea, I moved camp about five miles further westward, to a locality called Kalaieha. This point is now used as a sheep station. The pasturage upon the slopes of Mauna Kea is very abundant and rich, but there is no water. At first it was a mystery to me how these animals could flourish with nothing to drink. It appears, however, that the fog is so abundant that a night rarely passes without more or less rain or a condensation of vapor sufficient to thoroughly saturate the grass, and the animals thus obtain sufficient moisture from the grasses they feed upon. They seem to thrive very well, and I have never heard of any serious loss arising from want of moisture.

Kalaieha is situated near the summit of the pass between Mauna Kea and Mauna Loa, at an altitude of about 6,900 feet. Both to the eastward and to the westward there is a very gentle slope towards the ocean, so gentle in fact that from here it appears to the eye like a broad level plain. The lavas from Mauna Loa have flooded it again and again, and are now outspread over a vast expanse in fields of black, ominous, naked aa. These lava floods stretch all the way up to the very base of Mauna Kea and find a sharp line of demarkation upon its lowest slopes. The base of Mauna Kea is well covered with soil and volcanic sand, giving life to an abundant herbage and no inconsiderable number of trees, thus offering a strong contrast to the desolation and blackness of the lava fields beyond. Around us are very many cinder cones, some of noble proportions, and from the summit of any one of them we may obtain an overlook of these Phlegrean fields. The sense of desolation which they awaken is exceedingly impressive. In the preceding chapter I have already mentioned how the descending lava streams from Mauna Loa spread out over wide areas when they strike the comparatively level platform below. It is often difficult to distinguish one field from another, so intimately are they blended together and so faint is the distinction of color. Only when some field of extreme recency has been spread out like that of 1881, disclosing a superlative blackness, is it possible to comprehend its full extent and individuality, by its contrast with fields a little older and just beginning to show the first effects of weathering. The entire prospect conveys to the mind the idea that these flows succeeded each other at very brief intervals and that all of them are of great magnitude. The portion of any *coulée* which is comprised in its course down the mountain slope invariably bears a small ratio in respect of mass to the quantity spread out upon the lower plain. Nor do these currents by any means stop always at the base of the mountain, but deflect sometimes to the eastward, sometimes to the westward, according to the slope of the land. They stretch onwards towards the sea for a distance of many miles, and not a few of them have entered the ocean. This was the case with the great eruption of 1859, which entered the sea upon the western coast of the island, while the last eruption of 1881 came within about a mile of the sea at Hilo upon the eastern coast.

Several days were spent at Kalaieha searching for varieties among the lavas and for such other facts of interest as might present themselves. Very little, however, was discovered. The lavas of Mauna Kea, especially around the base of the mountain, show but little variety, and those of Mauna Loa are even more homogeneous.

Leaving Kalaieha, my next objective point was the valley of Waimea, on the northern side of Mauna Kea. To reach it it was necessary to go over the mountain. This was not a serious undertaking, for it presents no difficulty except the length of the journey, and this is readily overcome by dividing up the march between two days. The mountain was crossed upon its western flank by an easy trail and our camp was pitched near the summit of the ridge. From this point a fine view of Mauna Loa and Hualalai is presented. The huge lava streams descending from Mauna Loa to the northwest between Mauna Kea and Hualalai are distinctly visible and present a most suggestive aspect. The best defined among them is the great flow of 1859, which is visible in all its extent, reaching from a point near the summit to the sea, a distance of about 35 miles. The interval between Mauna Kea and Hualalai, which, reckoned from base to base, is about twelve miles, has been traversed by a great number of such lava floods within a very recent period in the history of the mountain. Viewed from a lofty standpoint on Mauna Kea, the general grouping of these beds and the long flowing profile which they have generated are presented to the eye most vividly. It is easy to imagine how, step by step and by flood after flood, this part of the island has been built up by the simple superposition of numberless lava streams.

Descending the northern slopes of Mauna Kea the plains of Waimea at length are reached. These plains are bounded by Kohala Mountain on the north and Mauna Kea on the south, and form a moderately elevated pass hardly 3,000 feet high between the east and west side of the island. The western declivity of this pass is arid, hot, and barren, suggesting the desert plains of Nevada. The comparison is strengthened by the occurrence of cacti, which seem to be very closely related to some of the opuntias of southern Nevada and Arizona, and the first impression is that they are merely varietal forms of the common prickly pear which have here attained a considerably larger size, but without any other change of habit. But the ubiquitous sage (Artemisia) is wholly wanting and seems to be about all that is needed to complete the similarity of the picture. In place of it are many low, sickly, stunted shrubs having the air and habit of desert plants quite as distinctively as the American sage. As we approach the summit of the pass there is a gradual but rather rapid increment and freshening of vegetable life. From the summit to the eastern coast the descending slope is clothed with abundant vegetation, which soon becomes a tropica jungle similar to that which we traversed in passing from Hilo to the base of Mauna Kea. Thus in the course of a very few miles the journey from west to east over the plains of Waimea will lead us from a region as truly desert as Nevada to a region where the ground is muddy by incessant fog and rain and incumbered with the densest of tropical forests. The cause of this extreme contrast is easily discerned. The perpetual trade wind striking the eastern coast is thrown upward nearly 3,000 feet in the course of about fifteen miles, and is depleted of a great portion of its moisture. It then descends as rapidly to the western coast, and of course becomes very dry. Through the Waimea pass a powerful breeze is always blowing from east to west. Its effects may be seen in many ways, some of which are sufficiently striking. All of the cinder cones, and there are many of them scattered around the base of Mauna Kea, are deformed, being built up more upon their western than upon their eastern sides. The steady wind has caught the showers of lapilli as they were projected upward and caused them to fall in much greater quantity upon the western sides, so that the vents are situated upon the eastern sides of the cones, giving them all a uniform aspect of deformity. The effect of the wind is also seen in the steady drift of the sand dunes, and even the clinkers scattered about upon the plains show a marked wearing upon their eastern sides by the ceaseless action of the sand blasts.

The little village of Waimea is situated upon the southern base of Kohala Mountain, a little west of the summit of the pass. It is a beautiful spot, seeming as we approach it from the south or from the west like an oasis in the desert. It lies just upon the verge where the arid region passes into the moist. A stream of delicious water, and perennial, comes down from Kohala Mountain, and flowing towards the western sea gradually sinks into the earth long before it reaches it. Like most other Hawaiian towns it is but a faded remnant of a population which was once considerable. There is still some thrift here, arising from rather exceptional advantages for pasturage. Curiously enough, horses and cattle seem to thrive best in a desert country when left to their own natural ways and devices. This is as true of these tropical islands as it is of Western America.

From Waimea we obtain a superb view of the northern flanks of Mauna Kea. As compared with the southern portion of that mountain there is one notable difference. This is in the amount of erosion, which is at once seen to be very much greater upon the northeastern or windward side. Several huge ravines are visible, commensurate in their proportions with the magnitude of the mountain. An observer viewing these gorges from the northern and eastern sides would be apt to conclude that a very long period of time has elapsed since eruptions of lava and cinders have ceased to exercise any appreciable effect in building up the mountain pile. Viewing it upon the opposite sides, he would be equally apt to infer a relatively brief period since the cessation of volcanic action. The difference in the effects of erosion upon the two sides is certainly very great; but I can hardly doubt that it may be fully accounted for by the difference in the precipitation alone. In noting the effect of atmospheric degradation upon the rocks of these islands, as well as in other countries, I have been most forcibly impressed at all times with the enormous disparity in the rates of weathering, where the only variable factor is the amount of atmospheric moisture. Wherever the climate is moist the lavas decompose with great rapidity, so that a very few years are sufficient to produce a very appreciable amount of superficial disintegration, and to start the vegetation growing upon the rocks. Wherever the climate is dry rocks of identical character—nay, even identical streams, passing from a wet to a dry region—preserve their freshness for probably a century or more. Many instances may be seen here of lava flows which descend through a belt of moisture to some of the dryest regions along the western coast (most notably in Kona), and as a general rule the portions which are situated in the moist region will simulate very great antiquity, while the portions in the arid belt upon the coast will look extremely recent. We should of course expect to find the degradation of rocks much greater in a wet locality than in a dry one, but the difference is considerably greater than might be at first supposed.

MAUNA KEA FROM THE SOUTH.

CHAPTER IX.

HAMAKUA—KOHALA—HUALALAI.

Waimea is a starting point from which a journey may easily be made to the eastern coast in the district of Hamakua. This district is certainly one of the most curious in the world, and indeed its features could barely exist elsewhere than in a tropical island situated within the trade-wind belt. The slopes of Mauna Kea, descending at first swiftly—afterwards more gently—toward the sea, are suddenly terminated on the coast by cliffs ranging from 100 or 200 to 700 or 800 feet in height. These coastwise cliffs are exceedingly steep, approaching in many places the vertical, and plunging at once into water of considerable depth. Here the coast receives the full brunt of the long steady swell of the Pacific, rolling for thousands of miles before the powerful trade-wind. The erosive action of the sea-waves is probably as efficient here as in any part of the world, for it is very powerful and unremitting. That the waves have eaten into this coast for several miles does not seem to be in the least degree doubtful.

The slopes leading inland from the crest of this cliff are for several miles very moderate, hardly exceeding 300 feet to the mile, suggesting a gently sloped and slightly diversified platform. This platform is deeply scored by a surprising number of large ravines, so abrupt that in many portions they may well be called cañons. Their sides have slopes varying from thirty to forty-five degrees, and have sometimes a greater angle of acclivity. The number of these is very great, and they occur at intervals averaging a little more than half a mile. Along the coast between Hilo and Kohala and within a distance of about forty miles there are nearly seventy of these ravines, reckoning both great and small. A land journey coastwise through this Hamakua district would be quite impossible, had not a very fair horse-trail been excavated with much labor and expense. This trail could not have been built along the margin of the sea because the deep water washes the face of a vertical cliff. It could not have been easily built far inland on account of the jungle and quagmire. There was no better way than to descend and ascend every one of these sixty or seventy ravines by zigzag courses dug out of the sides of the ravine walls. The only difficulty is the great labor and exhaustion which they entail upon the pack-animals. A man can go easily up and down on foot leading his horse without the slightest sense of insecurity and with no more fatigue than would be incident to ascending and descending a very long series of stairways, but the pack-animal must carry his burden.

The Hamakua district presents a very interesting juxtaposition of two very different methods of erosion. First, the surface and stream erosion of the land by rain-water; and, second, the erosion of the coast by powerful wave action. Not many years have elapsed since geologists were in the habit of attaching far more importance to the action of waves upon coast lines than to the steady degradation of land surface by those complex processes which pass under the name of denudation. We have only to recall the many discussions in England of the denudations of the Weald, and especially Charles Darwin's chapter upon that subject in his Origin of Species, to realize how strong was the impression only thirty years ago. At present the action of waves upon the coast is regarded as only a small factor in those processes which lead to the destruction of lands. And the geologist who views the windward coasts of these islands is tempted to raise the question whether geological opinion on this subject has not drifted too far to the other extreme and fallen into the habit of underestimating it. Not only in Hamakua, but upon all of the windward coasts of these islands, are the effects of wave action very extensive. It is not easy to form a very accurate estimate of the extent to which the Hamakua coast has been eaten away by the waves. But the probabilities are that the sea has made inroads varying from two to three miles upon the lowest flanks of Mauna Kea, and perhaps to a still greater extent upon the slopes of Kohala Mountain further to the northward. It is probable, also, that the period of time in which these inroads were made has been, geologically speaking, very brief, extending back presumably no further than the epoch at which the activity of these volcanoes terminated. So long as they continued to pour out their lavas and send their streams into the sea, so long we may regard them as building up the land quite as fast and even faster than the sea could destroy it. This is abundantly shown by recalling the condition of the coasts which bound the descending slopes of Kilauea and Mauna Loa. The lava streams from these active volcanoes reach to the sea, and, taking periods of a thousand or a few thousands of years, they not only repair the damages which the sea inflicts upon the coast line, but they even extend the land further out into the waters. We may, indeed, find numberless localities (in fact it is so with the greater part of the coast) where the sea, for a time, begins its inroad and makes some progress towards the formation of cliffs; but sooner or later a new lava stream overflows the line of attack, not only replacing all that has been eaten away, but adding more material besides.

Upon the coasts which bound the masses of Kilauea and Mauna Loa the cliffs are never high, seldom exceeding 50 or 60 feet, and along most or the coast lines are much less even than that. But the activity of Mauna Kea is evidently of some antiquity, and the damages inflicted upon the shore line by the ocean have gone without repair ever since the volcano became quiet. Kohala Mountain has been inactive for a still longer period than Mauna Kea, and the cliffs along its shore line reach a still greater elevation, and have probably suffered a greater amount of inroad and destruction by the sea.

Returning to Waimea we may use it as a convenient starting point from which to visit Kohala Mountain. This is the smallest of the five great volcanic piles which constitute the island of Hawaii, and is also the most ancient. But in speaking of its antiquity, I wish to be understood as referring only to the epoch at which its activity ceased, and not to the epoch at which it began. Regarding the initial epochs of these volcanoes we know nothing. Like all the other volcanoes of these islands its lavas belong to the basaltic group. Still they have peculiarities which serve to distinguish them in some measure from those of Mauna Kea and Mauna Loa. They appear to be less ferruginous and much more feldspathic. Olivin does not form, as a rule, so important a part of the mineral contents of the lavas, though in some of them it is fairly abundant. The group taken as a whole shows a tendency to approach the andesites. Extreme cases may be found in which it is difficult, apart from the general relations, to say whether they should be classed among the augitic andesites or among the normal basalts. Those intensely black, dense, and highly ferruginous lavas which are universal on Mauna Loa and Kilauea do not occur on the Kohala Mountain, so far as I have discovered.

The altitude of this old volcano is about 5,600 feet, and its general form and structure is very similar to that of Mauna Kea. It has many cinder cones, some of which are in an excellent state of preservation, while others are much degraded. The whole mountain is covered with a deep, rich soil, and the rock exposures are good only in the ravines which have been scored by the streams. The summit of the mountain is difficult of access on account of the boggy nature of the ground and the dense growth of vegetation. Large peat-bogs are formed there at the elevation of fully a mile above the sea. The occurrence of peat-bogs at such heights is by no means uncommon in the other islands, for they are found upon West Maui, and especially upon the island of Kauai. Upon the summits of the latter island these bogs are of very great extent. Their formation is, no doubt, dependent upon the fact that summits between 4,000 and 7,000 feet of altitude are usually buried in clouds.

Leaving Waimea, we may now proceed to the nearest point upon the western coast of the island. A good wagon-road leads directly down the slope of the Waimea plains to a little village called Kawaihae. Only a mile and a half from Waimea the climate has become quite arid, and the country presents the appearance of a desert.

We reach the coast at Kawaihae, eleven miles distant from Waimea, and then turn southward along its margin. The coast is low, shelving gently beneath the waters with only a slight ripple in its profile. Being under the lee of the island, there is no surf, but only a gentle swell, caused by the diffusion of the swell of the Pacific around the northern end of the island. Near Kawaihae stands a large heiau or sacrificial temple built by Kamehameha I, and the last one erected on these islands by the Hawaiian race. It is a more ponderous structure than the average of them, having higher and thicker walls, and is planted upon a rock overlooking the sea at an altitude of some 70 or 80 feet.

At a distance of about five miles from Kawaihae we reach a walled inclosure containing grass houses and a large cocoanut grove. It is a pleasant spot, clothed with maniania grass, and contains a pool of slightly brackish water by no means agreeable to the taste, but quite fit for use. Here we encamp for the night, pitching our tent beneath the trees, and are kindly cared for by some old natives, who bring us water, milk, and cocoanuts.

Starting early in the morning, we move onward still along the coast. At a little distance from the shore are rough lava fields somewhat grimy with age, and partially drifted over with sand and dust blown down and washed down from the uplands. Although we are apparently upon the slopes of Mauna Kea, it is soon evident that these lavas have come from Mauna Loa, having lapped around the flanks of Mauna Kea since the latter volcano became silent. As we move onward these lavas become fresher in appearance and more rugged. At length the trail deflects away from the coast, stretching obliquely inland, and leads us to the border of a field of aa. As we gain the summit of it, it is seen to be a most formidable one, for it stretches almost as far as the eye can reach, and is rough, horrid, and bristly to the last degree. But an excellent trail has been built across it, reaching for about four miles as straight as a dart, and the traveling is reasonably good. The lava is brimful of olivin, and seemingly basic in the extreme. Very plainly it has emanated from Mauna Loa, and has poured down through the intervale which separates Mauna Kea from Hualalai. We now realize the fact that this intervale has given passage to a great number of immense streams, thirty to forty miles in length. Mauna Kea has proved a barrier to these lavas upon the north, and Hualalai has proved a similar barrier upon the west, the two piles deflecting a considerable part of the extravasations of Mauna Loa into one broad channel or passageway leading northwestward from the summit of the mountain.

After three or four hours of steady travel the eye is caught by some notable topographical features situated inland about three miles from the shore. Very plainly they are terraces of the same nature as those which we observed at Hilea in Kau. While all other older topographical features hereabout have been mantled over and buried out of sight by the great lava streams from Mauna Loa, these terraces, which are apparently fragments of more extensive formations now concealed, have escaped burial. Even these have been partially overshot by branches and filaments of the great *coulée* of 1859. But they are not sufficiently overflowed to obscure at all their real nature. These terraces are three in number, the highest being about 1,400 feet above the

CLIFFS ON THE WINDWARD COAST.

KAMEHAMEHA'S TEMPLE.

level of the sea, and the lowest about 1,000 feet. It is another indication of the upheaval of the island, but the amount of uplift here appears to be only about half as great as we found it to be in Kau. It is rendered more conspicuous by a considerable fault, probably of quite recent occurrence, which has dislocated the terraces along a line nearly perpendicular to their fronts and extending inland.

The flow of 1859 enters the sea hard by. The lava is here pahoehoe of a normal type, and owing to the aridity of the climate is as fresh and clean as when it first cooled. It extended the coast of the island about three-fourths of a mile into the sea, with a front nearly two miles long.

Almost immediately after leaving the flow of 1859, we enter upon lava fields which have emanated from Hualalai. These are of more ancient date for the most part than those emanating from Mauna Loa, which we have just crossed. They also exhibit a difference in texture which is extremely difficult to define, though it is apparent enough to the eye. The trail now turns away from the seacoast, and, leading obliquely up the slopes of the mountain, at length enters upon a lava field which is evidently of very recent date. In truth it was erupted in 1805. Though far inferior in magnitude to the historic eruptions of Mauna Loa, it is by no means an inconsiderable one. It broke out at a point upon the northwestern flank of Hualalai at an elevation of nearly 4,000 feet above the sea, and reached the ocean in a stream about a mile and a half in width and about seven miles long. The lava is extremely ferruginous and contains considerable olivin and no conspicuous feldspar.

All the way from Kawaihae the trail has led through a region which is extremely arid and hot, and over lava fields which for the most part are wholly destitute of soil and occurring as the roughest form of aa. But as we mount the slopes of Hualalai we pass by a slow transition into a moister climate. Behind us the only trees which grow are the cocoanut and pandanus, beside the lagoons of the sea-coast. Now, at length, the trees once more make their appearance and gradually become more and more abundant, and with them are numberless shrubs, plants, and grasses suitable to a climate which is intermediate between moist and dry. After a hard day's march, we camp on the western flank of Hualalai, at the height of about 2,300 feet above the sea. A little below us is a curious-looking crater, which erupted in 1811, sending a stream of lava into the ocean. It is a jagged-looking cone, having the abnormal and abortive appearance which we have frequently observed around the orifices of eruption on Mauna Loa. Here is the point where the last sign of activity in Hualalai manifested itself. From that day to this Hualalai has remained in perfect repose.

Of Hualalai there is little that needs to be said. Its summit is about 8,600 feet high, and altogether it is a large volcano, though dwarfed by comparison with its giant neighbors. Its mass is built up like that of Mauna Kea, partly of fragmental material, and it has many cinder cones. Its summit is crowned with many of these cones, thickly clustered, and even confluent with each other, and all in perfect preservation. Hundreds them also stand upon the flanks of the main pile at all altitudes. There is, however, a somewhat more marked tendency than in Mauna Kea for the cinder cones to increase both in size and number towards the summit. The terminal ones are certainly the largest, though many large ones are seen lower down. Some of these cones have no cups in their summits, but the chimneys open with sharp edges at the mouths of the pipes from which the profiles slope away at once and continuously to their bases. The same structure is observable in the cones which stand within the caldera of Haleakala. The mean profiles of the mountain have about the same acclivity as those of Mauna Kea, and the ascent is easily made from the northwestern flank.

The lavas of Hualalai are all basaltic, containing much olivin and highly ferruginous. They resemble those of Mauna Loa in many respects, though with a slight difference, which is apparent enough to the eye, but eluding accurate description.

CHAPTER X.

KONA.—ERUPTION OF 1868.

The journey southward through the district of Kona, though an arduous one, is in many respects delightful and instructive. An old road, which is in reality nothing more than a horse trail, has been constructed along mountain slopes at heights varying from a few hundred to two thousand feet above the sea, and from two to four miles from the coast. It was formerly much used, but of late years, owing to the increase in the facilities of transit by sea and the decrease of the population, it is seldom traveled and is much overgrown by vegetation. South of Hualalai the western slopes of that mountain, and also of Mauna Loa, are covered with a dense tropical forest hardly so close and impenetrable as those upon the windward side of the island, but still very luxuriant. The climate is very rainy, and this fact may be put in contrast with the extremely arid climate prevailing upon the western coast of the island north of Hualalai. Why should one part of the leeward coast be very wet while another part of the same coast is very dry? The problem is an interesting one and the solution of it instructive.

As we ascended Mauna Loa and Mauna Kea, we had occasion to note that after reaching an altitude of a little over 7,000 feet the trade-winds were no longer felt. At a higher altitude we encountered the anti-trades. From the summits of those lofty piles, looking in every direction as far as the eye could reach, we observed that the aspect and movements of the clouds indicated that this fact was in no degree dependent upon the existence of the island itself or its mountain barriers, but prevailed everywhere within the limits of vision and presumably to an indefinite distance beyond. The trade-winds very plainly are a movement of the lower stratum of the atmosphere alone. In this neighborhood the thickness of this stratum is roughly a little more than 7,000 feet. I am assured by intelligent observers upon the islands that during at least nine months of the year and a greater part of the remaining three months of midwinter, the cloud-play over the island, and the observed motion of the winds are uniformly the same as I observed it to be. Wherever, then, the altitude of the land masses rises above this limit of 7,000 feet the trade-winds cannot blow over them, but are deflected around them. Looking now to the Waimea plains, which have an altitude of about 3,000 feet, the trade-winds blow over the summit pass in great volume, and descend upon the leeward slopes. As they mount the eastern slopes they throw down their moisture in great abundance, and by an inverse action of causes they become very dry as they descend the western slopes. But the district of Kona is situated under the lee of the lands which rise far above the limit of the trades, and is in no manner whatsoever influenced by them. The winds of the Kona district are quite at liberty to follow the daily alternation of land and sea breeze in their simplest and most typical mode of action. In the morning the sky is clear and the sun shines gloriously. Soon after sunrise the air is dead calm, but about 10 o'clock the sea breeze sets in blowing from the west and ascends the mountain slopes. Quickly the clouds gather, and at length the rain falls steadily throughout the afternoon and well into the night. At 9 or 10 o'clock in the evening the wind gradually ceases and soon after the land breeze prevails. Descending the long mountain slope the air becomes dry, the clouds clear away, the stars shine out, and the latter part of the night is cloudless. This alternation repeats itself with rare exceptions daily throughout the year. Seldom does a day pass in the Kona district without a rainy afternoon. And yet as a rule there is no rain at the lowest levels upon the margin of the sea. There is a narrow belt of land close to the ocean varying in width from a mile to a mile and a half, or even two miles, where rain seldom falls. Here the slope is comparatively gentle, and as the sea breeze blows inwards the effect of the ascending current is not felt until the stronger slopes a couple of miles away from the sea are encountered. At altitudes of 400 or 500 feet vegetation is abundant, and increases in density to heights of nearly 4,000 feet. Very notable effects upon the amount of precipitation may be detected from even a very slight change of elevation. This no doubt results from the fact that the winds from the ocean when they first strike the land are very near the point of saturation.

It is by no means necessary that the altitude of any land barrier in these islands should be as high as 7,000 feet in order to form a well-defined lee. In truth, 1,500 feet, or even perhaps 1,000 feet, produces a marked effect; but where the barrier is so low as this there is no alternation of land and sea breeze, for the trade-winds which blow at night as well as by day pour continuously over the lower barriers, and the climate under their lee is more or less arid. Only when the barrier is high enough to completely oppose the flow of the trade-wind is the daily sea breeze possible, and when such a sea breeze occurs upon the western side of the island it brings with it its complement of moisture and throws it down as the air ascends almost as copiously as the ascending trade-winds upon the windward sides.

Along the Kona road the tropical vegetation becomes exquisitely beautiful. Kona is the home of the kukui or candle-nut tree, of the bread-fruit, and the banana. Lower down near the sea, where the climate is dryer, the pineapple grows in wonderful profusion. The coffee tree grows abundantly, running wild over the entire district. Formerly it was much cultivated, and the Kona coffee is fully equal to the very choicest article that comes into our market, surpassing, in my estimation, the best government Java, and equaled only by the Liberian. The production some years ago was considerable, but the trees were attacked with blight, and the increase in the price of wages nearly destroyed the culture of the plant, so that now the production scarcely exceeds the very small domestic consumption. Many abandoned coffee plantations were seen and myriads of healthy-looking coffee trees scattered throughout the forest laden with flowers and with green, red, and ripe berries. It seems as if the superlative excellence of this fruit might promise some future revival of the industry.

For three days our journey lay through a country where every turn of the road opened visions of paradise. Nor does the vegetation ever become monotonous, for it is constantly changing in its habits and is full of variety. The district was once very populous. On every hill-side may be seen stone walls now fallen in ruins which once marked the subdivisions of the little tenements of the old Hawaiian agricultural population. The land appears to have been subdivided much more minutely than the lands ordinarily are in civilized countries. The number of these little tenements must have been vast indeed, and if we were to use them as a measure of population we might infer that the people numbered many tens of thousands. But such an estimate would be inadmissible, because there is no doubt that single individuals often held more than one tenement. And moreover it is probable that the place having been occupied for a period was abandoned or forcibly depopulated, and that the inhabitants migrated to some other place. Still the population was once no doubt quite dense. At the present day it is scanty in the extreme, probably numbering only as many scores as it did thousands a century ago. The general aspect is that of a country once cultivated but long since left to solitude and overrun with untamed vegetation. The orange, the lemon, the banana, the mango, the citron now stand everywhere wild and uncared for, while the coffee trees and guavas form thickets well-nigh impenetrable.

Passing the slopes of Hualalai we come once more upon the mighty flanks of Mauna Loa. Numberless great lava streams may be detected, many of them of very recent date, descending to the sea. Along the line of the road the vegetation is so exuberant and grows with such amazing rapidity that lava streams which are probably but a few centuries old are covered with soil and overgrown with jungle. But as they reach down to the sea-shore, where the climate is more arid, they disclose their individuality. There are also points of interest upon the sea-shore. In several places the shore has been invaded by the sea eroding into cliffs which now and then attain some notable heights. This erosion has been effected surely by wave action, and it is worthy of note that this is upon the lee coast of the island, where there is no heavy swell of the ocean, as there is upon the windward side. Yet the daily land breeze is always sufficient to set the sea to dashing upon the coast, and the swell of the Pacific from north and south circles around the island and is quite appreciable upon its long lee shore. It is to be borne in mind, also, that the waves of the ocean are extremely complex. No wave is ever really lost upon the ocean so long as it has sea room to travel in. It is no doubt obscured in the case of a heavy prevailing swell, such as that which is rolled up by the trades. But winds blowing thousands of miles away generate waves which travel on and are never lost. Hence we always find upon the calmest day under a lee shore that the waves are still breaking and throwing up their spray, even though there is no local wind which could generate them. Doubtless they come from vast distances. We need not be at a loss, then, to explain the fact that wave action goes on to no insignificant extent upon the lee shore of the island of Hawaii.

At Kealakeakua, the famous spot where Captain Cook was killed, there is a cliff, the height of which I could not judge accurately, but suppose it to be about 400 feet. This is rather exceptional, for most of the coast line shelves down to the sea easily, and the shore wall is seldom more than 20 or 30 feet high. Over this cliff in Kealakeakua Bay several lava streams from Mauna Loa have cascaded into the sea.

It is also noteworthy that at the distance of about a mile and a half or two miles from the shore the slope is usually quite gentle. But farther inland it rises with a strong acclivity and is sometimes extremely steep. I have suspected that during the gradual elevation of the island these abrupt slopes formed the coast and were eroded by the waves. Most assuredly these slopes are not such as arise from the normal arrangement of a vast series of flowing lavas, piled sheet upon sheet, and they call for some special explanation of their abruptness. I can venture, however, to do nothing more than offer a conjecture upon this point. These slopes are now covered with many recent lava streams, which have in a great measure masked their abruptness.

As we approach the southern limit of the Kona district the climate gradually becomes still more moist. The rains begin earlier in the day and continue further into the night. The road is exceedingly rocky and is greatly incumbered with Hilo grass, which grows here with a rankness exceeding, if possible, its rankness at Hilo itself. There is the same alternation of pahoehoe and aa, but the rocks are generally covered with a film of soil and overgrown with forest. The lavas are all of the monotonous type peculiar to Mauna Loa. But all that is visible of them merely attests the general fact of innumerable large lava streams descending from far up the mountain to the sea and piled over one another sheet by sheet. At length we pass the southern Kona boundary and once more enter the district of Kau. Again the climate changes, suddenly becoming dryer, and the trail enters upon a vast stretch of recent lavas, forming a belt about fourteen miles in width and consisting mostly of aa. It is by no means a single stream, but many. All of them are very young, but a great portion of them are old enough to have allowed the growth of large trees, although no soil has formed. Here, as in Puna, vegetation takes root among the clinkers almost as readily as upon the soil itself. Some of the streams, however, are so recent that little vegetation exists upon them. It is an impressive sight to behold this wilderness of lava extending as far as the eye can reach up the mountain side on the one hand and sloping down into the sea on the other. Where the lava reaches the ocean a considerable number of secondary cinder cones have been thrown up in the same way as we observed them in Nanawale, in Puna. There is no way of determining the number of individual *coulées* represented in this belt, but they all seem to suggest that they were poured out in rather rapid succession with intervals of only a few years between. If the degree of weathering is any criterion, it might be a fair inference that some fifteen or twenty large *coulées* had shot down through this belt within the last three or four centuries.

Through these clinker fields a trail has been built in a rigorously straight line without the least attempt to avoid a single obstacle. A journey of fourteen miles across this belt of clinkers brings us to the beautiful ranch Kakuku, where we are most hospitably entertained and enjoy a rest, looking upon the illimitable expanse of the Pacific. A journey of five or six miles over a comfortable road brings us back to our original starting point at Waiohinu. But before we leave this spot to complete the journey it is well to look at the traces of one of the most memorable recent eruptions of Mauna Loa disclosed right here at Kahuku—the eruption of 1868.

The Kahuku ranch stands upon the summit of a cliff or steep ledge extending up and down the slope of the mountain and increasing in height seawards. At the foot of this cliff runs the principal branch of the lava flow. Back of the crest, about a quarter of a mile, runs anothar branch, and a little beyond that a third. The ranch occupies a beautiful spot of green surrounded by black fields of lava, now only fifteen years old. The eruption of 1868 was by no means one of the largest of the historic outbreaks of Mauna Loa. On the contrary, it is one of the smallest, though in itself it is of grand proportions. But it is especially interesting because it was better observed than any other, and also because the circumstances attending it were exceptional. The other historic eruptions have all broken out from and flowed through localities seldom visited by white men—from the awful solitudes of the upper dome and through the middle zones occupied by the forest jungle and the desolate phlegræan fields of the interior of the island. The eruption of 1868 broke out in a region peopled by intelligent white men, though sparsely, and within two hours' ride of many more. The other eruptions have come without warning, like a thief in the night, and progressed silently without tremor or violence. The eruption of 1868 occurred at a time when the southern part of the island was rocked for weeks by earthquakes of an appalling character, such as humanity happily is seldom called upon to witness or to endure.

Before sunrise on the morning of the 27th of March, 1868, people residing upon the northwestern and western parts of the island observed a great cloud of smoke or vapor suddenly shoot upwards from the summit of Mauna Loa to an immense altitude, illuminated by the glare of extensive fires beneath. After continuing to ascend for about an hour it was observed that smoke or vapor ascended from several points below, along a line stretching southwestward from the summit. Soon after sunrise the entire mountain became obscured by the ordinary trade-wind clouds, and the whole scene was shut out from view. The following night became clear, but, when the clouds had left, no trace of volcanic action was visible. Early in the day following (March 28) began a series of earthquakes, which gradually became more frequent, and lasted for a period of two weeks. The number of shocks could be reckoned only by thousands, the most violent of which occurred on the 2d of April. For hours together the earth was in a constant tremor, with now and then a shock of exceptional violence. At about 3 o'clock p. m. on the 2d of April a prodigious earthquake shook the southern part of the island with terrible violence, and was felt with considerable force throughout the entire group of islands. Houses were overthrown or shaken down in ruins. Men and beasts were thrown upwards and prostrated. Trees swayed to and fro like reeds in the wind, and a series of waves traversed the land, the earth opening in wide cracks on the crest of the wave and closing together in the trough. It was at this time that the great mud-flow already described took place at Kapapala. The southern coast of the island sank from two to eight feet in different places along an extent of nearly 60 miles. A mighty wave rolling in from the sea—its crest reaching above the cocoa-nut trees upon the coast, and sweeping inland to a distance varying from half a mile to two miles—carried everything before it. In the space of a very few minutes eighty human beings, in a very scantily inhabited country, met a violent death. Hundreds of animals perished, and every structure was shaken from its foundations. Still the shocks continued with great frequency until the 8th of April, or a week after the great shock occurred. It was not until the 7th of April, however, that the great eruption took place. About 7 o'clock in the evening a great column of fire suddenly shot upward upon the southwestern slope of Mauna Loa, at a point situated about 3,700 feet above the sea. In a very short time the air was thick with vapor, which glowed with an intense light derived from the great fountains of lava beneath. These fountains were situated along a fissure, occupying about a mile of its length and interrupted only by short intervals. One great sheet of the fiery liquid, judging from present appearances, must have been upwards of 2,000 feet in length, ranging up and down the mountain upon the brink of the cliff, which has already been described. The lava was poured forth with immense rapidity and in enormous volumes. In a little over two hours it had reached the sea, from 10 to 11 miles distant. On its way it spread out into numerous streams, the largest of which lay at the base of the long faulted cliff already spoken of. At some parts of its course the lava ran with a velocity which was estimated by some of the spectators to have exceeded 15 miles per hour, and the estimated velocity is probably not too great. The streams at present cover a space from 2 to 3 miles in width, including, however, between them several narrow places which were not overflowed. Most of the lava is pahoehoe, though here and there are some considerable stretches of aa, notably at the terminations of the several branches of the flow. All of it is olivinitic to an extreme, and individual specimens can be selected containing large grains or nodules of olivin, which constitute more than half the mass of the lava.

The duration of this outbreak was exceptionally short, for it lasted only about four days. The quantity of material ejected was many times smaller than that from any of the eruptions of 1855, 1859, or 1881, although it was, no doubt, many times greater than the largest eruption which was ever known to come from Vesuvius.

This eruption also seems to have been attended with a greater amount of explosive violence than any other of which we have record. Apart from the earthquake shocks which preceded and followed it, an unusual quantity of vaporous products appears to have been given off. The air was filled with Pele's hair and volcanic dust, and great quantities of that exceedingly light basaltic pumice which is often found both at Kilauea and upon the summit of Mauna Loa were scattered far and wide over the country. This pumice is so light that it may be carried to great distances by the wind. Much of it was carried out to sea and was afterwards observed floating upon the ocean.

Following the main lava stream down to the sea, it spreads out into a wide field of aa. It is also pushed out into the ocean for the distance of quite half a mile, extending the area of the island just so much. At the end of the lava flow are three cinder cones, the origin of which is very curious, interesting, and suggestive. It appears that these cinder cones were formed by the contact of the lava with water. Their structure and general appearance are quite normal, corresponding in all respects with the cinder cones which are formed over ordinary volcanic vents. Here is an unquestionable instance of the formation of an ordinary volcanic crater by the adventitious contact of liquid lava with water. Nor are these cones by any means exceptional occurrences on this island. At the end of the great flow of 1840 from Kilauea, which strikes the ocean at Nanawale, in Puna, near the eastern angle of the island, three cinder cones were formed in precisely the same manner. Upon the northwestern base of Hualalai the eruption of 1801 produced a similar cinder cone at the water's edge. Along the western coast of the island, between Kailua and the southwestern cape, a distance of 60 miles, may be seen a considerable number of small cinder cones standing upon the verge of the land, which from their appearance lead to the inference that they were formed in precisely the same way.

From Kahuku to Waiohinu, the point at which our journey commenced, the distance is only 7 miles, and a well-built road, comparatively speaking, connects the two places. Near this road, and right upon the verge of the faulted cliff, are three large pits having an exact resemblance to those which we have noted in the vicinity of Kilauea—for example Kilauea-iki and Poli-o-keawe. So near the brink of the cliff are these pits that one of them has its wall broken away on the side nearest to the cliff, leaving a gap which opens upon a desolate field of lava at the base of the main wall. These pits are from 600 to 1,000 feet in diameter and a little over 300 feet in depth. They appear to be very nearly circular and symmetrical in shape, but their origin can only be conjectured. They are, no doubt, quite ancient, as their sides are covered with soil, talus, and expose rocks of considerable antiquity.

THE ERUPTION OF 1868.

CHAPTER XII.

MAUI.

The island of Maui lies northwest of Hawaii. The channel which separates them has a width of 28 miles. The length of Maui is 50 and and its largest transverse diameter 32 miles. Its total area is about 900 square miles. It consists of two lofty mountain masses separated by a narrow isthmus which at its highest point is only about 160 feet above the sea. The depression of the land by 200 feet would convert it into two islands. The eastern mass of Maui is considerably larger than the western, and is also much higher. It is wholly occupied by what seems to be a single volcanic pile. The name of this mountain is Haleakala. It is interpreted as meaning the house of the sun.[1]

The general form and structure of Haleakala are very similar to those of Mauna Kea and Mauna Loa. It has the same dome-like contour, and is apparently built in the same way, by the accumulation of lavas mingled with fragmental products. It has numerous cinder cones upon all parts of its surface, and though these are quite normal in form, none of them attain the large proportions of those seen upon Mauna Kea. But by far the most striking feature of this mountain is seen upon its summit. The upper portion of the mountain contains a caldera suggestive of the same origin and mode of formation as that which we have attributed to Kilauea and Mokuaweoweo, but many times greater in extent. A fuller description of it, however, will be given a little further on.

The western part of Maui also consists of a lofty pile of volcanic matter, reaching an altitude of about 5,800 feet above the sea, but the contrast which it offers to Haleakala is very great. East Maui (Haleakala) is a volcano, the activity of which has ceased at a very recent period, so recent, in fact, that we may wonder why no tradition of its activity has been preserved in the legends of the people. But no such tradition has yet been brought to light, and we must infer that several centuries have elapsed since its last eruption, while the appearance of some of its lavas indicate that those centuries cannot be many. West

[1] Some of the white residents, learned in the native language, suggest that this name should be Hele-o-ka-lá, which means the trap in which the sun was caught. *Hale* means a house, but *hele* means a trap. The prepositions *a* and *o* both signify *of*, but the former implies an active relation of the *la*, or sun, while the latter implies a passive relation; that is to say, a-ka-la means that the sun did something—perhaps built the house or dwelt in it. But o-ka-la means that something was done to the sun. Now there is a well-known myth that Maui, the great hero and Ulysses of the Hawaiians, laid a snare for the sun and caught him, compelling him to make the daylight twelve hours long instead of eight.

Maui, on the other hand, has surely witnessed no volcanic outbreak within any period which even in the geological sense can be called recent. Once, no doubt, it was a great volcanic pile similar to Kohala mountain, though much larger; perhaps similar to Mauna Kea, though much smaller. But it has been greatly ravaged by erosion. The entire mass has been sawed up in to fragments by numberless gorges and ravines, which have eaten the heart completely out of it. These gorges are of an extraordinary, not to say unique, character. From the margin of the sea or from the plain of the isthmus and upon all sides they recede far back into the interior of the mountain, enclosed by walls rising thousands of feet and wholly inaccessible to human foot. The scenery in these gorges is probably as impressive and picturesque as in the wildest glens of Norway or Switzerland and will fall not very far below Yosemite. They are all the work of erosion, and the material excavated to form them probably constituted more than half the original bulk of the mountain. All this means time, and the work has been done since the eruptive vents became silent.

The traveler who desires to visit Haleakala may choose indifferently from two steamers which leave Honolulu, one landing him at Maalaea Bay, on the south side of the isthmus, the other landing at Kahului on the north side of the isthmus. Perhaps the more frequent route is the former one. After landing at Maalaea he would cross the isthmus by a smooth road leading to Wailuku, a distance of about six miles. This is a pretty village situated at the opening of the famous Wailuku Valley, one of those magnificent gorges excavated in the mountain mass of West Maui. Two or three miles further on is the village of Kahului, situated at the head of the bay of the same name, upon the northern side of the isthmus. A little railway of very narrow gauge and traversed by baby locomotives now connects Wailuku with Kahului. The railway also extends 3 miles further eastward, to the sugar-mills of the great plantation of Spreckelsville, by far the largest plantation in the islands. The reclamation of these lands from the condition of a desert to one of fertility was a triumph of energy and enterprise. The soil of this isthmus is fine and deep, but the climate is arid. In its original condition it was hardly to be expected that this soil could ever prove very fertile, and for two reasons. In the first place, it has the appearance of a red ocher, and contains from 30 to 40 per cent. (!) of red oxide of iron. In the second place, the climate is very dry throughout the greater part of the year, and there are very few crops which demand nearly so much moisture as the sugar-cane. Arid lands, however, usually become extremely fertile when irrigated, but there was no running water in the neighborhood, excepting a few small streams at distances varying from 3 to 10 miles which were already utilized for the same purpose. The nearest available supply was found in a stream flowing down the eastern flank of Haleakala, about 25 miles distant. To bring the water to the plantation it was necessary for the aqueduct to cross or circumvent many cañon-like gorges and rugged spurs similar to those in the Hamakua district of Hawaii. The engineering difficulties were unique and of so grave a character that they perplexed very sorely the engineers who constructed the canals which supply the hydraulic mines of the Sierra Nevada. But the work, though enormously expensive, was successful, and a supply of water was obtained capable of irrigating about 2,700 acres. The crop of sugar obtained from this land was 9,000 tons.

On the road from Wailuku to Kahului are several interesting points which are sure to catch the eye of the geologist. A little east of the village and upon its outskirts rises a ledge of rocks consisting of consolidated coral sand. The sand grains which compose it are rather coarse and large and identical in nature with the white sand occurring along the beach to-day. It lies in a very thick stratum which is strongly and intricately cross-bedded. The observer who is not thoroughly familiarized with the phenomenon of cross-bedding would be apt to mistake the false bedding for the true. The true stratification of the mass has a dip a very little south of east or towards Kahului, and ascends directly towards the mountain mass of West Maui. This dip is variable, ranging from three to six or seven degrees. Some portions of this coral limestone, for such it is, have an altitude of very nearly 200 feet above the sea. It contains an abundance of fossils, all of which, so far as observed, are existing species. Its recent origin is beyond question. This implies an elevation within recent times of the mass constituting West Maui. It also implies that this elevation has not been shared by East Maui, for the strata dip towards that mountain as if their prolongations would pass under it, although we can hardly suppose in reality that they would extend so far.

There are other facts which in conjunction with this also testify to the recent upheaval of West Maui to the extent of quite 200 feet and perhaps more. As we came from Maalaea Bay to Wailuku we could not fail to be impressed with the very striking appearance of the alluvial cones formed at the opening of every mountain gorge; and of these gorges there are many. Every one of these alluvial cones has been deeply cut and trenched by the stream which it carries. This is most conspicuous in the large cone at the opening of the Wailuku Valley, which carries the largest stream of all. Not only has the trench been cut in the alluvial cone, but it has also been cut deeply through the coral limestone some distance away from the mountain. We have, then, the following facts. The mind recurs to a period during which these alluvial cones were gradually formed in the usual manner. Subsequently, from some cause which we must explain, the river ceased to build up the cone and began to cut into it and destroy it. Why should it have ceased to build and begun to destroy its own work? The explanation is a little complicated, but our studies in the mountain regions of Western America have disclosed to us the law and demonstrated it by thousands of well-attested examples.

Every mountain stream brings down large quantities of fragmental material in the form of bowlders, cobble-stones, gravel, sand, and clay, derived from the gradual disintegration of the walls and slopes of the amphitheaters in which it heads. The quantity and coarseness of the fragmental matter which a stream of given dimensions can carry is regulated by the velocity of the stream, and the velocity is regulated by the slope or declivity of the stream's bed. But this declivity varies along different parts of the course of the stream. The farther up we go, on the average, the greater becomes the slope. As it flows out towards the plain the slope steadily, though somewhat irregularly, diminishes. But as the declivity decreases so also does the velocity of the stream and its power to transport sediment. But when the velocity of the stream is slackened, then the sediment which it could carry easily along its swifter portions is at length deposited. Thus in the steeper and swifter portions of the course the bed of the stream is largely in bare rock or among big bowlders, while in the more sluggish portions beyond it is in shingle, sand, and clay. The stream, therefore, cuts down its bed in the steeper portions where the rocks are bare and builds it up along those portions where the declivity is small. In a word, the stream naturally tends to equalize its declivity throughout. At the base of a sharp mountain range is a point where the declivity suddenly changes from very great to very small—changes from where the tendency is to cut down to where the tendency is to build up. Here, then, at the opening of the mountain gorge the stream throws down a great part of its load and gradually builds up an alluvial cone. But now let us suppose the mountain range to be hoisted in a manner similar to that which has happened at West Maui. Here the declivity is increased all along the line, the power of the stream to transport sediment is increased, and instead of throwing it down it begins to pick it up again and gradually cuts into the alluvial cone and begins to form a new cone further out on the plain. This is just what has happened in West Maui. Upon the eastern side of its great mountain front every one of these alluvial cones has been scored in just this way. Placing this phenomenon in juxtaposition with the occurrence of these elevated coral limestones, the evidence of upheaval is thus seen to be cumulative, and, indeed, irresistible. Passing around the southern flanks of the cañons of West Maui, as far as Lahaina, we find this phenomenon to be persistent everywhere, and although it becomes less and less pronounced, on account of the increasing aridity of the climate in that direction, it is still universal and unmistakable. We shall find further on evidence that East Maui has not shared in this uplift. On the contrary, I think that we shall find some evidences leading to the presumption that a recent subsidence has taken place there, though these evidences may not prove to be absolutely conclusive.

Proceeding eastward from Spreckelsville, the road leads in the vicinity of the coast and through a country which includes some of the finest sugar plantations of the islands. The soil is exceedingly rich, and the yield of sugar is very large. But as this region is situated between wind and lee the rainfall is not sufficient for the excessive demands of the sugar-cane, and irrigation is therefore resorted to. The scenery hereabout is very picturesque, embracing in its foregrounds the gentle slopes of the volcano covered with the rich verdure of maturing crops. On the one hand hard by is the deep blue Pacific with its billows breaking upon the rock-bound coast. Above us rises the vast dome of Haleakala, and in the opposite direction are the rugged peaks and serrated spurs of West Maui. Very beautiful, too, is the crescent shore line of Kahului Bay, curving off its beach of white coral sand in exquisite perspective. Nothing is wanting to complete the picture. Earth, sea, and sky, the mountains, the plains, and the meadows are full of beauty.

As we move along the coast we observe increasing indications of wave erosion. At first we see a little ledge, a few feet in height, of black lava carved by the ceaseless dashing of waves. As we move onward this ledge increases in height. The gentle slopes of the mountain here terminate upon the brink of a low cliff, which increases in altitude further on precisely as we saw it upon the Hilo coast of Hawaii. As the coast trends around to the southward to make the circuit of the mountain it becomes exposed more and more directly to the incessant roll of the surf which comes from the eastward. So, too, we may presume, have the effects of erosion been correspondingly greater and the recession of the cliffs has proceeded inland further as the directness of this exposure increases.

At length we cross an abrupt ravine coming down from the slopes of the mountain, and as the eye follows it upward it is seen to become deeper and cañon-like. It is the first of a series of similar gorges, which become larger and more frequent as we penetrate to the windward and rainier side of the island. A good wagon road has been excavated up and down the sides of this ravine, and at one point, where the bank has been cut away to make the road, there is a phenomenon worth noting. It discloses the progressive disintegration of the massive sheets of lava. The results are very curious, though by no means novel, for others have observed them. When a massive lava sheet cools its contraction cracks it into fragments by numberless vertical joints. The percolating waters carrying the reagents of decomposition penetrate these cracks and gradually disintegrate the rock. Here it may be clearly seen that each fragment is attacked upon its surface and the decomposition slowly penetrates inward. After a time each fragment begins to show the concentric arrangement. The inner parts of the block become round or spheroidal, and the decaying portions take the form of shells, like the coats of an onion. At a certain stage of decomposition a vertical ledge of lava thus decaying seems to be full of rounded onion-like masses inclosed in a matrix of clayey soil. Ultimately these spheroidal con-

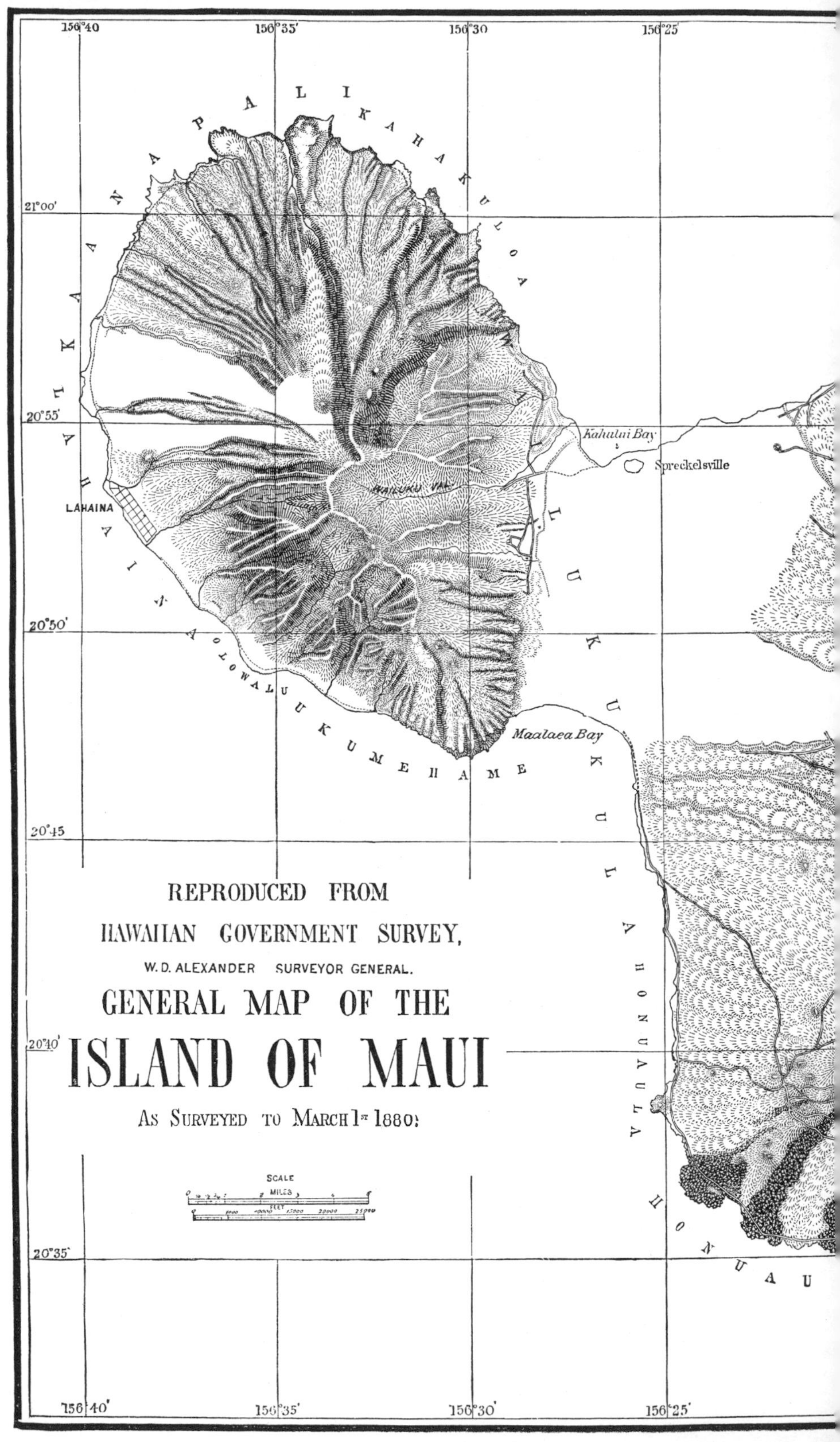
156°40
156°35'
156°30
156°25'
21°00'
20°55'
20°50'
20°45
20°40'
20°35'
KAANAPALI
KAHAKULOA
WAILUKU
KULA
LAHAINA
OLOWALU
UKUMEHAME
HONUAULA
Kahului Bay
Spreckelsville
Maalaea Bay
WAILUKU VAL
REPRODUCED FROM
HAWAIIAN GOVERNMENT SURVEY,
W. D. ALEXANDER SURVEYOR GENERAL.
GENERAL MAP OF THE
ISLAND OF MAUI
AS SURVEYED TO MARCH 1ST 1880.
SCALE
MILES
FEET
156°40'
156°35'
156°30'
156°25'

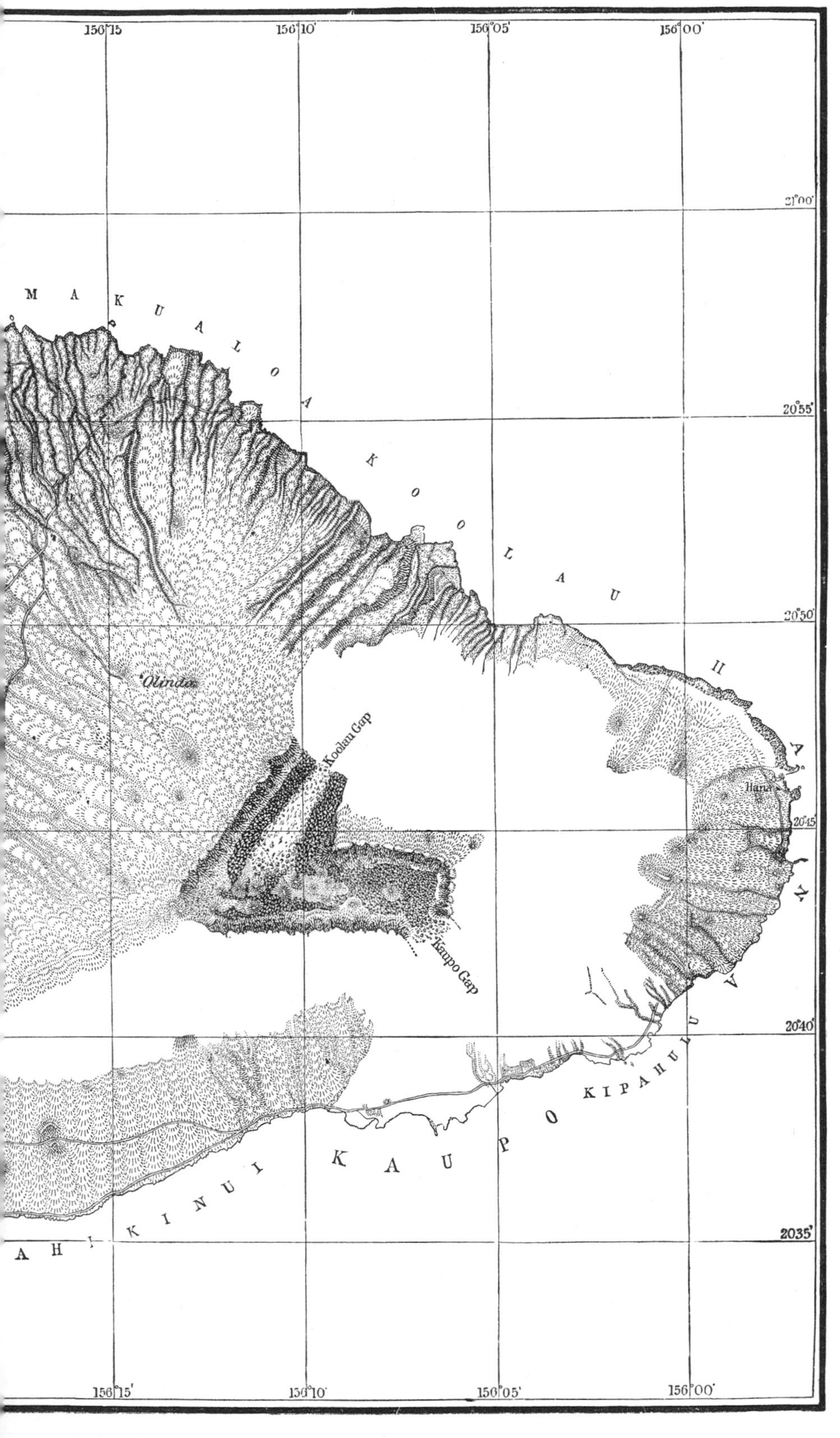
156°15'
156°10'
156°05'
156°00'
21°00'
20°55'
20°50'
20°45'
20°40'
20°35'
M A K
P U A L O A
K O O L A U
Olinda
Koolau Gap
Kaupo Gap
Hana
H A N A
K I P A H U L U
K A U P O
A H I K I N U I
156°15'
156°10'
156°05'
156°00'

WAILU

VIEW

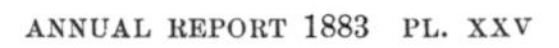

AUI.

EAKALA.

PUU NIANIAU
PUU PAHU
CAVE
CAMP
WHITE HILL
HAUPAKEA

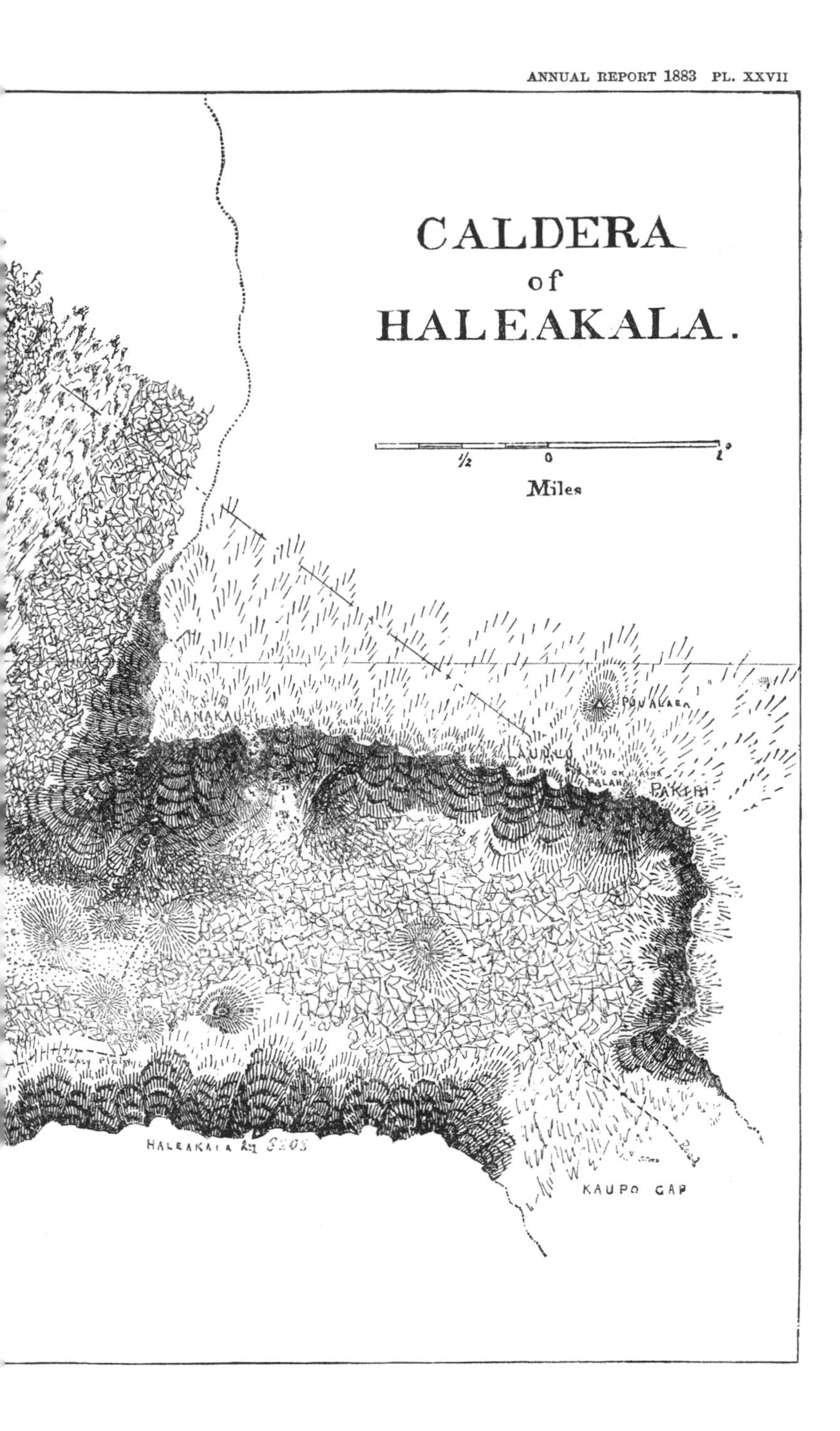
CALDERA
of
HALEAKALA.
½
0
1
Miles
HANAKAUHI
PUU ALAEA
PAKIHI
Grassy Plain
HALEAKALA
KAUPO GAP

centric shells also disappear, but so long as they remain visible they usually contain in the center a residual nucleus of undecomposed lava. In this locality, however, may be seen every stage of the process, beginning with the lava which has just begun to show the action of the reagents at the separate joints, and progressing through stages in which these fragments become more and more rounded and the enveloping shells more and more pronounced, and ending in a bed of clayey soil where all appearance of structure has vanished completely. This same phenomenon is disclosed very frequently in all of these islands wherever the exposed lava beds are of sufficient age. I have no doubt that these decaying fragments have often been mistaken for so-called volcanic bombs, and have been described as such. Here, however, the process is so clearly displayed in its various stages that it is not to be mistaken.

Half a mile further on we reach the plantation of Haiku. Here I was most hospitably entertained by the courteous proprietor, from whose house I started to begin the ascent of Haleakala. For a distance of about ten miles a good wagon road leads gradually up the gentle slope. It winds through a country which is fertile and picturesque in the extreme, and which enjoys a climate which few other spots on earth can rival either for comfort or salubrity. The temperature is nearly constant throughout the year, is never hot and never cold. The ordinary ranges are from 60° to 75° F. The rainfall is moderate and ample for almost every crop. The prospects are beautiful in the extreme. Of all places that I have seen or read of, none approaches more nearly to my conception of paradise than this.

The road ascends slowly until an altitude of 2,000 feet is reached. Here the ascent becomes much more rapid. The road dwindles to a mere trail, and the trail itself becomes gradually fainter. At the height of 4,000 feet we halted for the night at a little summer house planted upon a lofty knoll overlooking the distant sea, the broad isthmus below, and with the mountains of West Maui in the background.

Leaving Olinda, a faint trail winds up to the summit. Though sometimes rough and rocky, it is nowhere very steep or difficult. A sturdy mule will easily carry a heavy rider to the top. Along the route may be seen numerous old cinder cones in an advanced stage of decay and overgrown with vegetation. The rocks are mostly covered with soil sustaining an abundance of grass and heather-like plants. As the summit is neared the vegetation steadily thins out, becoming very meager, and at last almost vanishing. We come upon the brink of the caldera very suddenly and without any premonition of its proximity. In an instant, as it were, a mighty cliff plunges down immediately before us, and the famous crater of Haleakala is disclosed in all its majesty. Of all the scenes presented in these islands it is by far the most sublime and impressive. Its grandeur and solemnity have often been described, but the descriptions have not been overwrought.

Imagine two precipitous walls from 1,500 to 2,000 feet in height and each seven or eight miles in length, meeting at an angle of about 70°. At the bases of these walls is a plain from three to five miles in width. Across the plain rises a walled promontory, whose escarpments, of equal height, confront the two walls first mentioned. From the floor of the plain below rise ten or twelve large cinder cones, perfect in their forms, and apparently untouched as yet by the ravages of time. The central orifices of these cones are distinctly visible. Around the base of each of them is spread out a field of black basalt, here becoming confluent with the lava sheets of adjoining cones, there stretching away in a long flowing stream which disappears in the distance. In the overpowering presence of the giant walls around them, these cones at first seem small, but the experienced eye soon realizes that they are of no mean proportions. A solemn stillness and an air of superlative desolation brood over the scene. Every detail and lineament of volcanic action looks as recent and fresh as if the fires were quenched and the thunder of the eruptions had ceased yesterday. No growth of tree or grass or shrubbery is visible within, though upon the summits of the opposing walls the forest trees grow densely. The power of the scene consists as much in its desolation as in its vastness.

In two directions, eastward and southward, this vista of a volcanic plain studded with cinder cones and streaked with black lava stretches off between Cyclopean walls and vanishes by descending the mountain slopes. The eastern passage is named the Koolau Gap. The southern passage is named the Kaupo Gap. The former descends upon the windward side of the island and resolves itself into a huge ravine, and becomes confounded with a medley of vast mountain gorges scoured by erosion and encumbered with an impenetrable forest jungle. The southern or Kaupo Gap descends to a dryer region between wind and lee, and the walls gradually dwindle until at last they vanish.

It is difficult to give any precise account of the dimensions of this extraordinary feature because its extent is not well defined. The walls which inclose it have their maximum altitude at the summit of the mountain or in the vicinity where the two passes which compose it meet in an angle. From this point in every direction and down either pass the walls fade out and disappear gradually. The only dimensions of which we can speak definitely are the depth and width. The highest point of the surrounding cliffs stands about 2,200 feet above the plain immediately beneath. The widest part measured from the angle or coign where the principal walls meet across to the projecting angle of the promontory upon the opposite side, and from crest to crest, is about five miles. The height of the largest cone within the gulf is about 760 feet, or one-third the extreme height of the surrounding walls. The cliffs are almost everywhere sharply defined except at the coign where an easy slope is found leading to the bottom of the abyss below. Elsewhere the ramparts are extremely difficult to descend, and are practicable in only a very few places.

The trail from Olinda reaches the crest of the wall a little more than two miles east of the coign, and in order to descend it is necessary to skirt along the brink until the coign is reached. Everywhere a similar view is presented of the gulf below, but as we reach the angle other features are added to the scene. Right here stands a large cinder cone which forms the apex of the mountain. Its height is about 300 feet. From its summit we may gain a magnificent view not only of the abyss below, but far away in the distance to the southeastward, of the domes of Mauna Loa, Mauna Kea, and Hualalai, projecting above the domain of the clouds. These loftier realms of air are wonderfully clear, and with a good field-glass there is no difficulty in making out whatsoever details they have to offer with as much precision as if they were only a dozen miles away. The height of this summit cone upon Haleakala is about 10,350 feet above the sea, and its altitude is sufficient to give us one of those inspiring views of the realms of cloud land, as seen from above, which we have hitherto spoken of in our visits to the still loftier mountains of Hawaii. Through the rifts in the clouds we may behold the sea below, which has a strange visionary aspect that seems full of mystery and desolation. Behind us are the jagged peaks and bristling spurs of West Maui peering above the clouds and offering a fine contrast to the rounded domes of distant Hawaii.

The orign of this imposing feature of the volcano may be suggested to us by recalling the facts we have observed at Kilanea and Mokuaweoweo. There can be but little doubt that this abyss has also been formed by the sinking in of the summit of the mountain. But we must not presume that it was formed suddenly by a stupendous collapse of the upper dome. More probably it was formed progressively and slowly by the successive dropping of slices of the wall one after another, thus gradually enlarging the area of sinkage. True, this is largely a matter of conjecture, but we may attach more importance to it than we otherwise would had we not the evidence of just such a mode of development in the other great calderas. The dimensions of the pit of Haleakala are several times greater than those of Mokuaweweo and it is more than twice as deep. But as the development of Mauna Loa goes on, its summit caldera will probably continue to enlarge, and perhaps deepen, and there is hardly any theoretical limit which can be fixed for its ultimate dimensions. The shape of the caldera of Haleakala is also markedly different from those of Kilauea and Mokuaweoweo. But it would be difficult to suggest any reason why such creations should have any definite and common shape whatsoever, and therefore there is no reason here for doubting the identity of its mode of origin.

Perhaps some geologists and most laymen might raise the inquiry, Why may not this feature be attributed to erosion with as much probability as to vertical displacements or faulting? The reply is that valleys or gorges formed in mountains by erosion have forms and features directly associated with their mode of origin and determined by rigorous mechanical laws. These forms and features do not appear at the summit of Haleakala. To go into full detail in a discussion of these essential features would carry us far beyond reasonable limits. We may, however, advert very briefly to one or two crucial points. Wherever a great valley or gorge is eroded in a large mountain mass the head of the valley forms an amphitheater or series of amphitheaters, with abrupt and precipitous ravines immediately beneath the peak. In general terms, as we follow such a ravine from the plains below upward towards the summit the grade of its bed becomes steeper to the very last. Again, where two or more mountain gorges descending on different sides of the cone reach far up towards the summit so that their upper portions are separated only by a narrow divide, then this divide will always be sharp and well preserved through all stages of erosion. These characteristics I believe to be universal, and I am unable to recall any exception. Nothing of the kind is seen in Haleakala. Its floor has no other diversifications than the cinder cones which have been built upon it and the thin fluent streams of lava which have issued from their vents. If these cinder cones were shoveled away the floor would be flat and smooth to a striking degree, and only gently declining by an almost imperceptible slope in the directions of the two gaps, one to the eastward and the other to the southward. Nowhere can we detect any of the evidences of an erosional origin. Erosion here has done nothing more than to make many small carvings and incisions upon the wall faces.

It seems necessary, however, to call attention to the fact that such an origin as we have inferred can be attributed only in very rare cases to the chasms, gorges, and amphitheaters of mountains. Not one in a hundred of such chasms has this origin. In non-volcanic mountains such features are never produced in this way, and it is rare to find them even in volcanoes. The gorges which are seen upon the windward side of Mauna Kea, and which we shall soon see on the windward side of Haleakala, are unquestionably due to erosion. This great summit crater is a rare exception.

The descent to the floor of the caldera is very easily effected here at the coign. A long slope leads downward, covered with fine lapilli and volcanic sand, into which the feet of the animals sink deeply. By zigzag courses the declivity may be made very easy and gentle. Reaching the plain below, all that is necessary to secure easy traveling is to avoid the fields of fresh lava which are generally found near the bases of the cinder cones. The eruptions of most recent date all appear to be of trivial volume, and contrast by their very insignificance with the mighty outpours of Mauna Loa. Here, too, may be seen admirable illustrations of the common fact that cinder cones are built after the lava has ceased to flow. The fresh sheets of basalt are clearly seen underlying the cones, which have evidently been built over them. Nowhere do we find indications of the lava running either out of the summit orifice of the cone or laterally from its flanks. So far as could be seen every one of these cones had its terminal orifice quite perfect. Some of them have no cup around the opening, but a sharp edge like the cones on Hualalai.

All of the recent basalts which are poured over the floor of this caldera are of the most normal type, and extremely fresh in appearance. As we look at the irregular edges of the lava sheets exposed in the surrounding walls, and scrutinize the fragments which have fallen from them, we are at first inclined to infer that they differ from the fresher fields below. They, too, are typically basaltic, but on the whole the differences do not seem to be any greater than such as may be reasonably attributed to those changes which time and weathering invariably produce upon all eruptive rocks. The lava beds in the walls are certainly thicker and more massive, and this fact might justify us in the inference that these closing eruptions of the volcano were more feeble and much less voluminous than those which occurred during the period of full activity in the building of the mountain. We also observe in the surrounding walls numerous fragmental layers consisting of tuff and volcanic sands or ashes. These, however, do not form nearly so large a proportion of the component masses as the solid lava sheets, and are relatively much less abundant and voluminous than similar fragmental accumulations in volcanoes of more violent type. There appears to be no way of instituting a comparison between Mauna Kea and Haleakala as to the relative proportions which the fragmental masses bear to the solid lavas. However, from a general impression which I should be at a loss to explain very fully, it seemed to me that the fragmental material was far less in Haleakala than in Mauna Kea, though by no means inconsiderable.

As we proceed along the base of the main wall of the amphitheater and study the details of its front we observe almost everywhere a persistent recurrence of a certain kind of jointed structure. This is produced by cleavage planes of which the most conspicuous are always parallel to the face of the wall, and which are inclined at an angle of about 70° from the horizontal. Subordinate planes of cleavage perpendicular to the face of the wall, and nearly or quite vertical, are also seen, but they are less frequent and not strongly accented. All of them are made more conspicuous by weathering. This jointed structure gives in many places a rather striking aspect to the face of the cliff, and is somewhat suggestive of architectural effects. The jointing seemed to me to be due to a shearing force just such as is frequently produced in the vicinity of a great fault. It is often observed upon the flanks of some of the great faulted monoclines of the Plateau country, and I infer that the cleavages were produced by the same cause which led to the dropping in of the mountain platform during the enlargement of the caldera.

The journey over the floor of the great pit is smooth and easy, except where we are obliged to cross the recent lava fields, and even then the difficulty is never great nor at all comparable to the passage of the frightful clinker fields of Mauna Loa. The chief difficulty attending a protracted visit here is the want of water. The rainfall appears to be somewhat scanty, though fog and mist are abundant, especially in the day time, and frequently at night. All visitors speak with enthusiasm of the majestic scene presented by the great banks of cloud which every morning roll up the Koolau Gap and diffuse themselves among the cinder cones, wrapping them in fleecy folds and chasing one another slowly in wide circles around the crags and into the recesses of the walls. Upon the summit of the mountain this magnificent cloud drama may be watched with delight, for only now and then do the vapors sweep up so high as to veil the loftiest points.

At the mouth of the Kaupo Gap the floor of the caldera gradually bends downward and acquires a steeper declivity towards the sea. Here we come upon larger and rougher fields of basalt which look quite recent, though obviously older than the extremely fresh basalts which are spread about the bases of the cinder cones. Most of them have the form of *aa*, but are not nearly so rough as the great fields of Mauna Loa. Here and there patches of soil have accumulated in the swales, mosses have overgrown the clinkers, grass and scrubby vegetation have taken root among them. Our camp in Haleakala was just at the opening of the Kaupo Gap, 7,600 feet above the sea, where the more rapid descent to the ocean begins.

Leaving camp early in the morning, we descended. It is no light task to go down more than 7,000 feet of slope over fields of rough clinker. The journey, however, was accomplished by noon, nearly all of it being done afoot, leading or driving the tired animals. Very little matter of special interest was presented by the way. The great walls of the pass steadily diminished in altitude downwards, but did not wholly vanish until they nearly reached the sea. Everywhere they are exceedingly abrupt and almost precipitous; and yet they had sufficient inclination to give a foothold to literally thousands of wild goats that seemed to swarm upon their sides and filled the air with the continuous murmur of their bleating.

Reaching the sea-coast, we halted an hour for rest and then moved onward parallel to the shore towards the east. Here is a well-built trail, without which travel would be impossible. The country in front of us is precisely similar in its features to the Hamakua coast of Hawaii. It ends upon the sea in a vertical cliff, while the platform is sawed by cañons descending from the mountains. As the cliffs plunge at once into deep water, it is obviously impracticable to proceed along the water line. There is no resource but to ascend and descend the walls of each ravine as it presents itself. So abrupt are the sides of these gorges that without a trail only the most experienced mountaineer with the best trained animals would ever think of encountering them. As it is, the trails are very well built and make the passage easy, at least to the rider, and inflict no other hardship upon the animal than such as must necessarily attend the frequent ascent and descent of steep hillsides. How many of these ravines we crossed during that afternoon I have quite forgotten, but I should judge they were about twenty-five. Most of them were from 400 to 700 feet in depth. The intervals between them were rarely so great as half a mile, and sometimes the separating platform was reduced to a mere edge, so that as soon as the trail reached the summit on one side it instantly descended upon the other.

But the beauty of these gulches is quite beyond description. Every one of them is a scene of tropical splendor which neither tongue nor pen can portray. Though all are extremely beautiful, there is one in particular which seems to surpass all the others. It is named the Waialua Valley. The surrounding walls, 500 to 600 feet high, are carved into pediments of fine form and overlaid with a vegetation so dense, rich, and elegant that the choicest green of our temperate zone is but the garb of poverty in comparison. Five streams of water in different places cascade in snowy veils hundreds of feet down the precipices. Inland, the ledges of verdure-clad rock rise to a great height, with a receding vista of similar scenes. Over all is an atmosphere filled with a languid haze that blends and softens all beneath. The plants which above all others give richness and exquisite texture to the scenery are the great tree-ferns. They grow in dense thickets, covering the valley slopes and hillsides in broad masses. Their fronds, four or five feet in length and of most elaborate patterns, produce at a little distance an effect ineffably rich and delicate.

Long after nightfall we rode up to a fine mansion where dwelt the proprietor of the Hana plantation and received memorable hospitality. It was sorely needed. We had descended that day from the caldera of Haleakala, 7,600 feet above us, and had ridden and walked 20 miles more up and down, I know not how many cañon walls. Men were weary, and brutes well-nigh used up.

The eastern extremity of Maui has a broad margin of low land between the mountain and the sea. A portion of it is covered with deep, rich soil, suitable for a large sugar plantation. Much of it is overflowed with lava, which is not yet broken down into soil, but which maintains a jungle of tropical forest like the clinker fields of Puna. For a distance of 10 or 12 miles no ravines occur, the land being too low and flat. Further eastward, however, it rises again considerably, and gulches deeper and wilder than before are encountered in quick succession. We left Hana on the afternoon following our arrival there, and camped near the first ravine. The next day was spent in crossing the ravines until about 2 o'clock, when the animals became exhausted and could go no further. We halted in one of the largest gorges, and were kindly cared for by natives residing there.

There are some indications that the eastern portion of Maui has undergone a moderate amount of subsidence at a very recent period, though the evidence which is here offered may be regarded as inconclusive by many. The gorges which open to the sea out of the bounding wall of the island appear to have been once cut more deeply than their present floors and to have been partially refilled with shingle brought down by the streams flowing in them. The sea now enters the mouths of these valleys and in some cases extends half a mile or more within their openings. In other parts of the islands, West Maui, for instance, or eastern Oahu, where we have unquestionable proof of a rise of the land, the case is very different. There alluvial cones or deltas, composed of detrital matter, are pushed out into the sea. But at the eastern end of Maui the condition of the gorges at their mouths is indicative of refilling by the deposit of alluvium within them, which is easily explained by the assumption that the shore has sunken recently. The subsidence, however, does not appear to have been great, and was probably not more than a hundred feet, an amount which would be ample to meet all the facts observed.

Owing to the exhausted condition of the animals, we engaged a party of natives to take us in a boat the remainder of the distance to our starting point at Haiku. Rising at midnight we embarked with all our baggage and reached the landing place in the midst of a drenching rain at daybreak.

My visit to Maui was far too brief for my own satisfaction. It would have been most interesting to visit the western side of Haleakala to see some very fresh-looking lava-fields, which are notable objects as the inter-island steamer passes by them on its way to Hawaii. They bear the appearance of great recency when seen from the deck of the vessel about a mile off-shore; but it must be remembered that they are situated under the lee of the mountain, and in a very dry climate, and our observations in Hawaii and elsewhere have taught us that in such a climate basaltic lavas may preserve a remarkably fresh aspect for centuries, while they may look very old after an exposure of a few years in a wet one. Nevertheless it seems highly probable that the last eruptions of Haleakala took place only a very few centuries ago. No tradition of such an event is known to me as having been preserved by the natives, but that is not at all surprising.

Equally great was my regret at being unable to visit thoroughly the mountains of West Maui. Their wonderfully picturesque features alone would have been a sufficient inducement. But to have visited all points of interest would have required many months, and the time at my disposal was very limited.

CHAPTER XIII.

OAHU.

The island of Oahu offers to the geologist many attractive features, but I shall confine myself to a brief treatment of such only as present matters of special interest.

Like Maui it consists of two mountainous masses separated by a low-lying but broad isthmus. The eastern portion is the larger, but the western portion contains the loftier summits. Its rocks are volcanic throughout, excepting a few raised beaches of consolidated coral sand. The eastern portion of the island is not an individualized mountain mass like Haleakala, but is a true volcanic range, presenting an axis along which many ancient vents are situated, which have poured forth great sheets of basaltic lava, accompanied by considerable quantities of fragmental products. The trend of this range is from southeast to northwest, but has a considerable curvature concave towards the northeast. From the fact that the crest of this range forms a considerable arc of a circle, some observers, deeply impressed with the magnitude of the so-called crater of Haleakala, have conjectured that it might be a portion of the rim of a ruined crater which was once many times more vast. A careful examination of the structure of this range most effectually dispels this idea. The best notion of its structure and composition will be obtained by viewing it upon its northeastern flanks. Here it fronts the sea and receives the full brunt of the trade-wind, with cliffs 2,000 feet high and of magnificent aspect. In the face of this cliff may be seen innumerable sheets of basaltic lava, with intercalary beds of volcanic conglomerate, which dip away from the axis of the range in a northeast direction towards the ocean. Upon the opposite side, which is under the lee of the mountain, there is a prevailing slope of considerable but variable inclination southwestward towards the broad isthmus and the plain below, upon which is situated the city of Honolulu. Many large and picturesque ravines are scored in the leeward side of the range, and in the abrupt slopes of these gorges may also be seen the edges of the lava beds and layers of conglomerate dipping to the southward or just in the opposite direction to the dips observed upon the windward side. Thus the arrangement of the sheets which compose the mountain mass is easily seen to be similar to that of the tiles on either side of the roof of a house. But the present topographical features of the two sides of the range offer an extreme contrast. The windward side presents a gigantic cliff which is hardly accessible to human foot except at one point where now a good road has been built with much labor and skill, and which descends the cliff by zigzags. On the leeward side the profile descends from the summit to the sea in a very moderate slope, and has the general form of an inclined platform deeply incised with large ravines.

Notwithstanding the very powerful contrast at present existing between the two sides of the range, I think it can be made clear that originally a great mountain slope existed upon the windward as well as upon the leeward side, and that the windward slope has been carved away by erosion.

As the observer enters the harbor of Honolulu and perceives the long sweeping slope of the mountains descending towards him and notes the great ravines which have been cut in it, he will have no difficulty in persuading himself that these ravines must have had the same origin as those which occur in every other mountainous country. The bedding of the lava is distinctly seen, and the broken edges of the sheets are easily recognizable at a distance fronting in the steep slopes of the valleys. The imagination can easily restore the continuity of those beds and replace the material which has been removed in the excavation of the gorges. Making a mental picture of this restoration, the mind will have before it a smooth mountain slope very similar to what is now seen upon the westward or leeward side of Haleakala or upon the slopes of Mauna Loa. But instead of looking back into past time and restoring the material which has been removed, let the imagination range into the future many thousands of years and endeavor to conceive what will be the result of the operation of the same forces and processes which are now active. As in every other mountain country, the ravines would grow wider, their sloping sides would be gradually pared away, and the rocks reduced by secular decay to sand and soil. The silt would be carried off by the running streams to the ocean, and the remnants of the sloping platforms between the ravines would grow narrower, until at length they were reduced to sharp knife-edges, and would still continue to dwindle in size. Only let the time be long enough, and let the process go on indefinitely just as it is going on to-day, and the range will at length disappear. I conceive that the action upon the windward side of the island has been precisely the same in kind as upon the leeward, only much greater in degree. This difference in degree is unquestionably due to the difference in precipitation. Upon the windward side the rainfall is heavy; upon the leeward side it is light. And no doubt ever since the mountain range was built this same relative difference in rainfall has prevailed. We should therefore expect to find a proportionate difference as between the two sides in the amount of havoc which erosion has wrought upon them respectively. Even upon the leeward side we can see that it is very great. Probably it would be a low estimate to infer that one-third of the original mass of the leeward slopes has already been removed. It can scarcely be doubted that the precipitation upon the windward side and the corresponding intensity of the denuding forces are certainly three or four times as great. Let us recall here the contrasted aspect of the two sides of Mauna Kea. Upon the one, erosion has already made vast inroads. Upon the other its effects are only just beginning to be notable. So, too, upon Haleakala; the windward side shows ravines of great depth and magnitude, while the leeward side is as yet almost untouched. Not only does the amount of erosion increase with the amount of rainfall, but observation teaches us that under circumstances similar to these the difference in result is more than proportionate to the difference in the amount of rainfall.

The windward side of Eastern Oahu has therefore, as I believe, been almost completely swept away by erosion. The work thus inferred is no doubt a very grand one, yet it is by no means unprecedented. On the contrary, we can point in other parts of the world to other instances so much more vast that this falls into insignificance by comparison. But the present case is interesting because it has apparently been accomplished within a period which in all probability has not been very long when viewed from a geological standpoint of time. It also presents some peculiar characteristics which are not exactly parallel to the results of erosion presented by other regions. So interesting are they that they will bear some analysis.

As we journey along the northeastern coast of the island following the road which winds along near the base of the great cliff we find ourselves upon a rugged and well diversified platform which descends by a gentle slope to the sea. We observe at intervals of two or three miles, and sometimes much less, irregular ridges and chains of hills stretching out in long lines perpendicularly from the base of the great wall and ending as long promontories pushing out into the ocean. All this is well delineated upon the map. Some of these long spurs or ridges expand laterally, and wherever they do so we find very large hills, terminating in jagged peaks and sharp needles, rising to great heights and almost large enough to be called mountains. At first there seems to be nothing but confusion in their arrangement, or apparent want of arrangement, but upon more careful examination it will be found that every one of them belongs to some one of a chain of hills which forms a range like a mountain chain in miniature. Still, again, let us consider the valleys. There are many of these, and all of them head in amphitheaters in the main mountain wall and extend outwards from its base towards the sea. These valleys in most cases ramify, having subordinate lateral valleys heading each in its own amphitheater.

The origin of this topography is, I think, not far to seek. These valleys were once ravines carved in a long mountain slope stretching from the summit of the main mountain range far to the windward into the sea. As time went on these ravines were enlarged and gradually widened. With the progress of erosion the portion of the platform which separated any two contiguous ravines became narrower and narrower on its top until at last it was reduced to an edge at one or more points. Still the degradation of the valley walls went on, the valley constantly growing wider and the separating barrier gradually dissolving, until nothing was left of it excepting these lateral spurs and chains of hills which we now see. This is precisely the way erosion operates in all mountainous countries, and the results which it accomplishes through long periods of time are very similar to what we see here.

The decay of the rocks is so slow to the senses that the mind instinctively regards its consequences as of trifling effect when viewed in relation to the general degradation of the country, and naturally repels the thought that its final result is the casting of mountains into the sea. And yet it is but the plainest deduction of reasoning that this very result of a process apparently so feeble must follow inexorably if the process be continued long enough. This throws us back upon the difficulty of conceiving the immensity of time which the geologist is obliged to invoke; and yet I apprehend that the first impression of this sort is apt to be much exaggerated. We might at first suppose that this process would have to be kept up for very many millions of years in order to achieve the results which are here ascribed to it. But this is not so. There is more danger of exaggerating the time than there is of underestimating it. In a country favorably situated for the efficient action of the erosive force a thousand years would produce, I believe, a very marked change in the aspect of the mountains and valleys. In ten thousand years the changes would, indeed, be very great; and in a hundred thousand years who shall say how vast the changes would be? It is most unfortunate that all efforts which have hitherto been made to measure absolutely the duration of geological time, or to measure the periods which are requisite to effect any important geological changes have proved futile. Still they may not be regarded as having been wholly barren of results. To assign a period sufficient for accomplishing erosion so great as that just now seen upon the windward side of Oahu is impossible. And yet it may be practicable to take maximum and minimum limits widely apart and say with some confidence that the true period will probably fall somewhere between the two. From what we know of erosion in other regions, and taking account of the various factors which affect the efficiency of erosion in this particular locality, I think it would be safe to affirm that the period of this erosion has not exceeded 500,000 years, and has not been less than 50,000. Though the value of such an estimate is not very great, it may not be absolutely worthless.[1]

---

[1] In Dr. A. Geikie's new Manual of Geology carefully revised estimates are given of the rates of denudation in the drainage basins of six rivers, viz: The Mississippi, Upper Ganges, Hoang-Ho, Rhone, Danube, and Po. These estimates are based upon a large mass of data obtained by a series of carefully made observations. They are no doubt very fair approximations to true present rates. From these it appears that the Upper Ganges is carrying away annually from its drainage basin an amount of

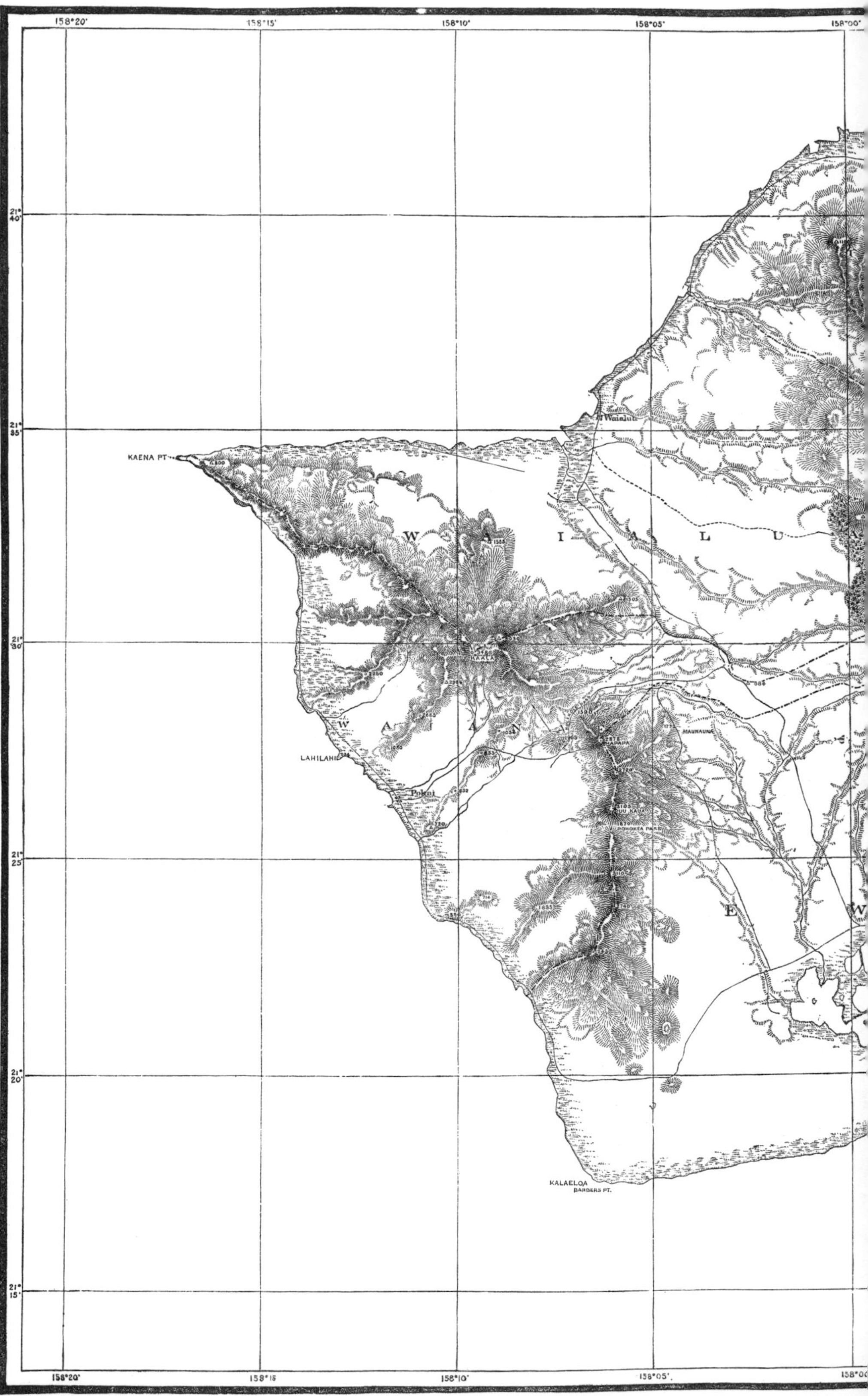

158°20'
158°15'
158°10'
158°05'
158°00'
21° 40'
21° 35'
21° 30'
21° 25'
21° 20'
21° 15'
KAENA PT.
Waialua
LAHILAHI
Pokai
POHOKEA PASS
KALAELOA
BARBERS PT.

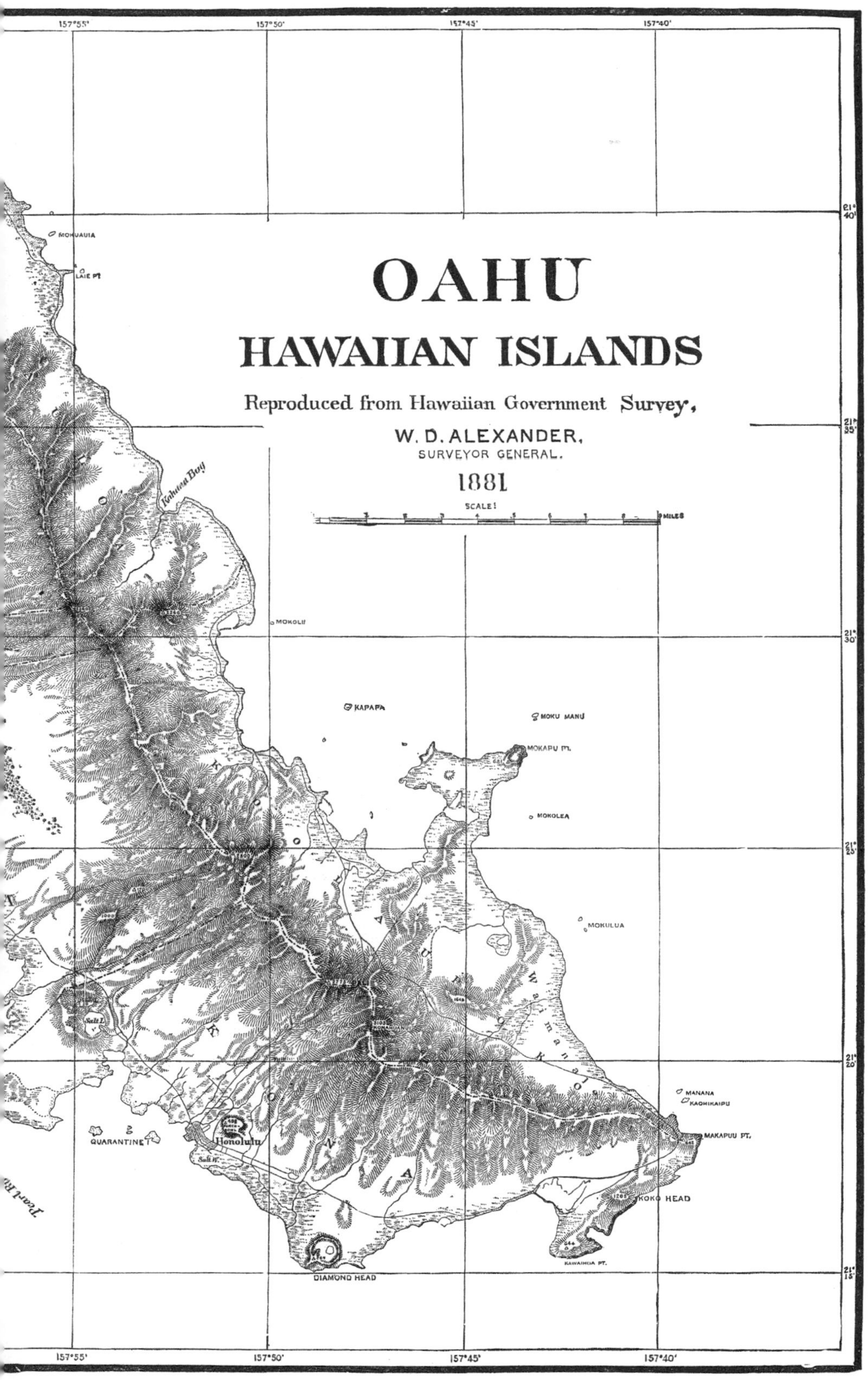

OAHU
HAWAIIAN ISLANDS
Reproduced from Hawaiian Government Survey,
W. D. ALEXANDER,
SURVEYOR GENERAL.
1881
SCALE
MILES
157°55'
157°50'
157°45'
157°40'
21° 40'
21° 35'
21° 30'
21° 25'
21° 20'
21° 15'
MOKUAUIA
LAIE PT
Kahana Bay
MOKOLII
KAPAPA
MOKU MANU
MOKAPU PT.
MOKOLEA
MOKULUA
Waimanalo
KONAHUANUI
MANANA
KAOHIKAIPU
MAKAPUU PT.
QUARANTINE
Honolulu
Salt L.
KOKO HEAD
KAWAIHOA PT.
DIAMOND HEAD

THE PALI—EASTERN SIDE OF OAHU.

DIAMOND HEAD.

The geologist who has had an opportunity to study erosion under the widely varying conditions presented throughout the United States may still find much of novelty in the results of that process disclosed by these islands. In our own country we have nearly every known variation in the quality and texture of the rocks in their modes of stratification, in the climate, and in the altitude, all of which conditions have their bearing upon the final result. But these tropical or semi-tropical islands present new phases. The most marked characteristics of the erosional forms here presented are their exceeding sharpness and abruptness and the development of great cliffs of erosion. I must confess to some surprise upon becoming conscious of this characteristic. Of all regions of which the topography is characterized by cliffs and crags, the most remarkable in the world is no doubt the Plateau country of the Rocky Mountain region which is drained by the Colorado River. I have elsewhere attempted to portray the symmetry and imposing character of these cliffs and of the innumerable cañons cut to wonderful depths in the cliff-bound platforms. But that region has a very arid climate. And the protracted study of it has left no doubt in the minds of our geologists that this aridity has played an important part in determining those peculiar forms. I was hardly prepared to find great cliffs and cañons in a country where the rainfall is excessive, and indeed so great that it is surpassed only in a very few localities throughout the world. But it is even so. It does not appear, however, that the conclusions drawn with respect to the effect of aridity upon the cliffs of the Plateau country are in any measure invalidated. We are only reminded more forcibly perhaps than ever before of the fact that similar forms may be produced by a variety of causes, and while under certain conditions the arid climate may favor their production, it is not always essential to it.

These very abrupt features are by no means peculiar to the Hawaiian Islands. They are seen also in the Society Islands, in the Fiji, and in fact throughout all of the lofty volcanic islands of the Pacific. It is by

material which, if spread uniformly over the entire basin, would have a thickness of $\frac{1}{823}$ of a foot. That is, at the present rate, the Upper Ganges carries off a mean thickness of one foot of material in 823 years from its entire water-shed. The Po similarly removes one foot in 729 years. Now, it can hardly be doubted that the average rate of denudation of the valley walls and slopes of the windward fronts of the Hawaiian Islands is greater—even much greater than in the Valley of the Po—and for two reasons: First, the rainfall is much heavier; second, the slopes of the valleys and their inclosing sides and walls are, on the average, very much greater than those of the Valley of the Po, and, from the nature of the circumstances, must have been so throughout the entire period of degradation. But (other things equal) the rate of degradation increases very rapidly with the declivities. The basaltic lavas and tufa beds are also quite as susceptible to decay as the average of sedimentary and metamorphic rocks, if not more so. Hence we may safely infer that these windward fronts waste away very much more rapidly than the basin of the Po. Therefore it seems to me that a very few hundred thousand years would suffice to effect the entire destruction which I have inferred for the windward side of Oahu.

no means probable, however, that these results are at all anomalous. They may without much difficulty be accounted for. The whole question turns upon the relative importance of the two groups of forces which together make up the total process of erosion; first, the efficiency of those forces which disintegrate and break up the massive rocks; second, those which gather up and carry off the *débris*. Wherever the former group bears an unusually larger ratio of efficiency with respect to the latter the sculptured forms will be mild and obtuse. Wherever the reverse is the case the forms will be sharp and abrupt. Under ordinary circumstances a moist climate favors disintegration more than the transportation of *débris*. Under such circumstances the *débris* accumulates in the form of very thick soils and immense masses of alluvia, which form a protecting mantle to the rocks and generate gentle slopes. The arid climate, on the other hand, no doubt diminishes both the disintegration and the transporting power of the streams. But the streams of a dry continent do not derive their waters from the low-lying, arid regions, but from distant mountain sources, and the detritus which has formed very slowly in the arid region itself is easily washed away by the showers, even though they be unfrequent, so that, upon the whole, disintegration is retarded more than transportation. The case of these islands is peculiar; even in the oldest of them the quantity of running water is not great. The rain which falls is absorbed by the lavas, cinders, and tufas with much more than usual facility, and sinking into the depths of the mountain masses probably makes its escape into the ocean beneath the level of the sea. The streams which do exist, however, have abnormally great declivities. Their power to corrade is therefore enormously increased and they cut down their beds at a very rapid rate. No doubt the weathering of the valley slopes and precipitous walls is here also exceptionally rapid. Still transportation and corrasion are favored much more than the disintegration of the rocks. That is to say, both processes are favored, but transportation and corrasion more than disintegration.

One of the most striking objects to be seen in the vicinity of Honolulu is a large pile situated three or four miles southeast of the city and known as Diamond Head. It is composed of cinders and tuff, and is, in fact, an immense cinder cone. Within it is a very large crater, more than a mile across. Its rim is a sharp edge which forms a complete circle, and though higher in some portions than in others, it is nowhere broken down. To enter the crater it is necessary to climb a steep slope and descend into it upon the other side of the crest. The highest point of this ridge has an altitude of about 760 feet. The outer flanks of the cone are scored upon all sides with little ravines, which give it the aspect often presented by the fronts of the Bad Land cliffs of Dakota or of the Plateau country. The cone is situated close to the sea, which washes the foot of its southern slope. As we pass around this flank we find a mass of strata composed of consolidated coral sand, which is strongly cross-bedded. The highest visible exposure of this coral rock is about 200 feet above the sea. That it formed once a wave-washed beach just beneath the surface of the ocean is self-evident. Thus we have a clear indication that this portion of the island has at no distant period been hoisted by at least that amount. As yet the waves have made no particular inroad upon the tufa cone itself, although atmospheric erosion has cut into it quite deeply, and upon all sides, the ravines already referred to. It seems very plain that this cone is of very much more recent date than the main masses which compose the mountains of eastern Oahu. This is inferred at once from the fact that the amount of degradation in the cone is utterly insignificant in comparison with the immense gorges which are seen even upon the leeward side of the mountain chain.

Diamond Head is a sample of a considerable number of similar cones which appear to have been formed at a period comparatively recent and very long subsequent to the principal period of volcanic activity during which the great mass of the island was built up. Immediately above the town of Honolulu stands another cinder cone very similar in its structure, which has received the name of the Punch-bowl. A little to the east of Diamond Head are several others, one of which stands upon the margin of the sea, which has demolished by wave action a large part of one side of it and exposed its interior structure. Still further eastward four or five miles is another large cone over a thousand feet in height, but less ample than Diamond Head in the area of its base. It forms a conspicuous object as the steamer from San Francisco approaches the island, and is one of the best known landmarks. It is named Koko Head. Upon the windward side of the island, at some little distance from the shore, and surrounded by the ocean, may be seen three or four other cinder cones which have been greatly ravaged by the sea, and which are now rapidly dissolving away under the remorseless action of the waves. There is still another remnant of recent volcanic action which deserves notice. The traveler who visits these islands, even though it be for the few hours allowed by the stoppage of the Australian steamer as it touches at this port, will most probably take a carriage to visit the Pali, as it is termed. A good carriage-road leads from Honolulu by a gentle ascent up the largest and broadest valley on the leeward side of the range until it reaches the crest of the main mountain wall. Here there suddenly breaks upon him one of the most beautiful and picturesque views in the world. He stands in a notch upon the brink of a majestic cliff carved in most sumptuous fashion and in the strangest forms, and looks down 1,300 or 1,400 feet upon the eroded platform below and upon the sea beyond. A zigzag trail of ample width has been hewn out of the cliff-side and winds down to its base. Descending this steep path there may be seen half way down, upon the right-hand side a steep slope, consisting of lapilli arranged in the regular fashion of cinder cones, evidently showered over and molded to the irregularities of the rock-faces which existed before the cinders were thrown out. It would seem that right upon the crest of the mountain range a volcano had broken forth within a comparatively recent period and long after the principal period of activity. Many other instances of recent cinder cones are to be seen in eastern Oahu, all of which appear to indicate a resumption of volcanic activity after a period of very long quiet. Relatively speaking, the results so produced were not great, and the activity was diffuse and comparatively feeble. It is interesting, however, as an illustration of a fact which has been observed in other regions, viz, that volcanic activity may cease for a long period in any locality and again resume operations. As a matter of history we know that this was the case with Vesuvius, which broke out with terrible energy in the year 79 A. D., after a period of repose so long that no tradition of an eruption from it existed. Similar evidences have been discovered in some of the more recent volcanic fields of the Rocky Mountain region. We may also advert here to the fact that the volcano of Hualalai has given no sign since the year 1811.

Appendix A
Dutton's Assignment to Hawai'i

William R. Halliday

After years of geological field work in the high plateaus of Utah and northern Arizona for the U.S. Geological Survey, Clarence Dutton was scheduled to head new field studies in the volcanic Cascade Mountains of Oregon and northern California in 1882. But on very short notice, Survey Director John Wesley Powell assigned him to Hawai'i for a field season lasting nearly 4 months. With travel by slow boat, this required about 6 months, the normal summer field season for Survey geologists. In his administrative report for 1882–1883 published in the same volume as "Hawaiian Volcanoes" (pp. 22–23), Dutton advanced several reasons for this assignment. At that time they apparently were acceptable to a (mostly) admiring Congress and administration, and to fellow geologists. In retrospect, however, they seem a trifle disingenuous. Soon after, Congress briefly prohibited the spending of any Geological Survey funds for fieldwork outside the United States.

Dutton began his explanation by stating that "inasmuch as the geologist must be preceded by the topographer," field geological studies in the Cascade Mountains would have been premature alongside a topographer crew in 1882. This does not ring entirely true. Dutton's previous fame had arisen from simultaneous topographic and geological studies in more difficult terrains than the Cascade Mountains.

Dutton continued with a much stronger reason, however:

> For [study of the Cascade Mountains] it was necessary that the observer should bring with him as much knowledge as possible, derived from the field investigations of corresponding phenomena in other regions. My own field studies in volcanism had been limited to the [very scant] facts presented in Utah and Arizona. It would most probably have been a well-founded criticism upon further work for anyone to suggest that this field had been entrusted to one who had never seen a live volcano....

This is certainly plausible, but a querulous critic could point out that many other studies of volcanic regions have been performed by geologists who have never seen an active volcano. Dutton continued:

> The choice of regions lay between Iceland and the Hawaiian Islands. Concerning the Mediterranean volcanoes much has been written, and information almost as good as actual examination can be obtained by reading. To a less extent, but still in a marked degree, the same consideration applies to Iceland, and, moreover, the difficulties of travel are so great in that island that it seemed doubtful whether much could be accomplished in the brief space of an Arctic summer.

Analysis of this seemingly plausible mix of assertions requires at least a little geological knowledge. The sentence about Mediterranean volcanoes virtually negates his earlier assertion about the need for actual observation. More to the point was the fact that the only dependably active Mediterranean volcano was (and is) little Stromboli, whose observation would have contributed little to studies of the Cascade Mountains. Vesuvius, Etna, and Vulcano erupt so infrequently that even a prolonged visit would not have ensured observation of their activity. Volcanoes of Iceland also erupt intermittently and unpredictably. There, too, he would have had to become familiar with travel by small boat. Supposed restriction of fieldwork to the "Arctic summer" of July and August, however, was a schoolboy's excuse. The month of September is an excellent time for geological studies on that subarctic island. On the other hand, Kīlauea and/or Mauna Loa volcanoes in Hawai'i had been erupting continuously for several years. It was the best choice.

Thus it appears that someone other than Geological Survey Director Powell initially suggested that Dutton go to Hawai'i, and that Dutton jumped at the chance. In view of circumstances that have not been considered by mainland-oriented biographers of Dutton, an additional agenda may have existed. If so, its roots go back to 1878.

In Washington in 1878, Dutton suggested to Powell that they form a Cosmos Club, "a center of good fellowship, a club that embraced the sciences and the arts, where members could meet socially and exchange ideas, where vitality could grow from the mixture of disciplines. . . ." Discussion was free, frank, and traditionally off the record. The club was spectacularly successful. Soon its membership encompassed most of the Washington Establishment: leaders of Congress, the administration, and the military as well as leaders in American science, technology, and intellect. From its onset, exploration, study, and conservation of natural resources were emphasized. Distinguished visitors soon were commonplace, from Rudyard Kipling to the Sierra Club's Dave Brower. Much of the world's subsequent history took shape around its polished tables.[1]

---

[1] The Cosmos Club now occupies an impressively stately building in the heart of Washington's Embassy Row. In addition to services to its own members, it provides meeting space for the Washington Group of the Explorers Club and related organizations.

During the club's first decade, the shaky relationship between the Hawaiian Kingdom and the United States undoubtedly was a frequent topic of conversation. King Kalākaua's triumphal 1874 visit to Washington would have been well remembered in 1878, along with the goodwill the "Merrie Monarch" had generated for his kingdom. The subsequent Reciprocity Treaty firmly aligned Hawaiʻi within the political sphere of the United States and briefly put Hawaiʻi's economy on a sound basis.

The islands' economy became increasingly cash-starved, however. Seemingly reckless actions arising out of astounding drinking bouts by the monarch were matched by surprising actions by the unpredictable legislature. Seeking to renew the goodwill generated by his 1874 tour, King Kalākaua surpassed that tour with a 10-month around-the-world extravaganza. But fate intervened. This time there would be no triumphal reception in Washington, D.C., no renewed bonhomie between the chiefs of two sovereign nations. While Kalākaua still was on the other side of the world, a shocked America went into mourning. President James Garfield was fatally wounded by a crazed office seeker. Until September 19, he vainly fought for his life.

Through the U.S. Minister to Hawaiʻi, the Washington establishment was receiving wild rumors about supposed threats to American investments—and supposed increasing Hawaiian support for annexation, an issue largely dormant since the death of King Kamehameha III. Others claimed that sharp American businessmen like Claus Spreckels were profiting hugely from weakness of the monarchy and the legislature. With King Kalākaua en route early in 1882, Cosmos Club pundits must have been sorely starved for reliable, objective information. Powell was a consummate insider in the Washington establishment and his geopolitical interests were no secret. But if anyone at the Cosmos Club (or in the National Academy of Political and Social Sciences) suggested Dutton as an ideal person to observe and report objectively, the secret never leaked. The text of "Hawaiian Volcanoes" yields only tantalizing hints. Its first chapter does reveal Dutton's sentiments: American investments were prospering, hence there was no reason for annexation to the United States. In the Maui chapter, he praised Spreckels' enterprise. That is all.

But something notably is missing at the end of the book. It is abruptly truncated, without the expectable synthesis of why "Hawaiian Volcanoes" was important to its readers. For a prize-winning author, only one explanation seems likely: a vigorously worded summary was deleted hastily as the manuscript went to press.

A summary, of course, might have dealt purely with geological matters; that science was wracked with bitter controversies and vigorous recriminations. But enlightened readers cognizant of the machinations of political factions in 19th-century America may be excused for suspecting that the missing summary contained anti-annexation sentiments unacceptable to someone at a higher level.

Probably we will never know the full reasons for Dutton's sudden assignment to Hawai'i, nor for the uncharacteristically abrupt ending of his book. But the outcome is clear. Annexation did not occur for another turbulent decade and a half, after Dutton had left Washington for San Antonio. For crucial years after 1882, Dutton remained a highly respected authority on Hawai'i in the heart of the Washington establishment. Clearly he merits recognition as an anti-annexation bulwark.

Appendix B
Biographical Sketch of Clarence Edward Dutton

William R. Halliday

Until the student of geopolitical unrest focuses on its subtly unusual aspects, Dutton's life seems transparently ordinary for his times: just another brilliant White Anglo-Saxon Protestant rising from poverty to professional fame and a touch of fortune through education and wartime bravery. In Dutton's case, however, it wasn't quite that simple.

Clarence Edward Dutton was born May 15, 1841, in Wallingford, Connecticut, and died January 4, 1912, at his son's home in Englewood, New Jersey. No adequate biography exists; his colleagues had difficulty obtaining even minimal information for his professional obituaries. His trunkful of personal papers disappeared from a storage warehouse, and even his own son did not know intriguing details of his remarkable life.[1]

The Dutton family was well-reputed in New England but his father is reported to have worked simultaneously as a shoemaker and as the poorly paid postmaster of a small town. This was hardly affluence, but the family clearly valued traditional education. One of his brothers became a lawyer; the other graduated from West Point and became a career military officer. As a boy, Clarence was intended for Yale University. At 15 he was considered ready to enter its divinity school but was held back a year. At 16 he matriculated, but after two weeks he transferred to Yale's incipient liberal arts program "before they threw me out" he later wrote, Mark Twain style.

At this time Yale College was heavily grounded in traditional classics, which taught self-discipline, produced enormous self-confidence in those who passed, and challenged student memories to ever more prodigious levels. Seemingly only an average student, his later life centered around these characteristics. By 1845, Yale also had begun requiring its seniors to hear lectures in chemistry, mineralogy, geology, astronomy, and other sciences. Dutton showed no special brilliance in geology. He especially excelled in mathematics and literature, and he surprised almost everyone by winning the Yale Literary Prize as a senior. He graduated at 19 with a further

[1]The best of the scant biographical notes about Dutton is Wallace Stegner's 1936 Ph.D. thesis.

reputation as a public speaker and an oarsman, but he remained at Yale for two additional years. Presumably this was for postgraduate study but perhaps for another reason: Two years into the Civil War he returned to Yale to marry a New Haven girl as soon as he received a noncombat billet. Of his subsequent family we know almost nothing, and little more of Dutton as a person; he abhorred publicity seekers to the point of almost pathological secrecy about himself. It is clear, however, that in later years he maintained his association with Yale, most notably with Professor James Dana, a member of the Silliman Dynasty that brought geological fame to Yale while Dutton was an undergraduate. Their professional relationships evidently were complex and have not received the analysis they merit. By the 1880s Dana was recognized as one of the world's leading geologists, but a letter from Dutton to Dana in 1883 seems more than a little condescending.[2]

During 1862 the Civil War escalated. Dutton entered the Union Army in September 1862 as 1st lieutenant in the 21st Connecticut Volunteers and adjutant to his brother who was its colonel. His brother was killed in action during the bloody Wilderness Campaign while acting as regimental commander, and Dutton himself was seriously wounded during the vicious battle of Fredericksburg. Promoted to captain, he remained with his regiment for a time, but he became fascinated by the mathematics of artillery. He passed a formidable examination, accepted a demotion, and became a career officer in the Ordnance Corps. In 1864 its assignments took him to Pittsburgh, Wheeling, Cincinnati, Springfield, and central Tennessee—a remarkable involuntary tour of the varied geomorphologies of the eastern United States. During the last year of the war, as a mere 2nd lieutenant he commanded the ordnance depot of the Army of the Potomac. This position almost certainly brought him to the attention of General and President-to-be Ulysses S. Grant and perhaps also Major John Wesley Powell. Powell had lost an arm during the war and was destined for fame not only as a geologist but as an unlikely explorer of the wild Colorado River and its canyons.

After the war, Dutton was assigned to the Watervliet Arsenal in West Troy, New York. Peacetime duties were light, and academia was nearby in Troy and Albany. Dutton's innate curiosity about ordnance initially led him into advanced studies in chemistry. But at the noted paleontological collection of the New York State Museum in Albany he encountered the study of invertebrate fossils. At that time these fossils were the key to dating sedimentary rocks, first as single beds and then entire sequences. Here he studied intensively with James Hall and other noted geologists who disagreed with Dana's theories on elevations and depressions of the Earth's surface. But his first scientific paper, in 1869, concerned the chemistry of the brand-new Bessemer method of steel production.

[2] Dutton's letter to Dana, 1883.

After five years at Watervliet, Dutton was transferred to the Frankford Arsenal in Philadelphia. Evidently he carried a cordial introduction to the long-established scientific community of that city. He was especially well received at the American Philosophical Society, the Philadelphia Academy of Sciences, and the Franklin Institute. At a meeting of the first, he soon presented an impressive geological paper on "Regional Elevations and Subsidences." Evidently he impressed someone in authority. After a mere year in Philadelphia he was transferred to the arsenal in Washington, D.C., and soon was promoted to captain of ordnance.

Through the Philosophical Society of Washington, Dutton quickly was welcomed into the company of distinguished leaders in many branches of science, government, and warfare. Especially notable was Major Powell, fresh from two expeditions down the turbulent Colorado River that had made him a national hero. War-amputee Powell undoubtedly felt a strong personal bond to the battle-scarred officer who could converse intelligently whenever the conversation turned to geology but who preferred to listen. The two soon became close friends. Together with field geologist G. K. Gilbert they formed an unofficial triumvirate of an emerging American school of geology based on exciting discoveries beyond the Great American Desert. A notable organizer and enormously fertile in ideas, Powell was something of a protégé of President Ulysses S. Grant. This evidently did no harm to Dutton's career and circle of friends. Curiously, however, he became a member of the National Academy of Political and Social Sciences as well as predictable geological associations.

Dutton and the American West

At the time of the American Civil War (1861–1865), the face of the American West was known very imperfectly. Little was on record about its geology, and a meaningful survey of its natural resources was virtually impossible until the general nature of the region was understood. A United States Geographical and Geological Survey of the Rocky Mountain Region was created soon after the war. Major Powell was its director and it thus became known as "the Powell Survey." Its exploring teams expanded scant prewar knowledge. Typically they traveled west each spring and returned to Washington each autumn to process their data and discuss their findings with other members of the Washington establishment.

Joseph Henry (then secretary of the Smithsonian Institution) was among those impressed with Dutton's comprehension of geology and obvious leadership skills. He and Powell persuaded Dutton, and then the Army's Ordnance Corps, that Dutton should be detailed from the Army for detached duty with the Powell Survey. This peculiar arrangement lasted 15 years—it had to be specifically approved by Congress four times and produced five of the most notable reports in American geological history.

Prior to this western fieldwork, Dutton's training in geology was entirely informal. But he learned rapidly from his new colleagues and from direct observation. To this he brought his earlier knowledge of chemistry and physics. A well-known 1877 report by G. K. Gilbert included a report by Dutton on unusual volcanic features of Utah's Henry Mountains.[3] Subsequently, volcanic occurrences became his primary interest. By 1880, he had studied enough local lava flows to conclude that volcanic loci were not mere funnels down into the molten core of the earth, as many had speculated. Instead, their products seemed to emerge from localized pockets that somehow were heated to the melting point of rocks, at much lesser depths. For 30 years he groped for the source of this volcanic heat.

At this time, geology was scarcely a unified body of knowledge. Bits and pieces of information still were being fitted together into comprehensive concepts. Numerous blind alleys still were being championed by distinguished professors. Major progress occurred in 1878 when the U.S. Geological Survey superseded the old "Powell Survey." A year later, Powell himself became director of the Geological Survey and Dutton took his place as chief of its Division of the Colorado. During summer field seasons he especially compared various models of mountain building with what he observed—especially in the rugged, unmapped high plateau country of south-central Utah.[4] He found it notable for a thick sequence of horizontal beds of rock, presented on an enormous scale. Such a pattern for a high plateau was especially inconsistent with one widely accepted theory of mountain building: lateral pressure caused by shrinkage of an aging earth. James Dana and many others then believed that mountains arose like ridges on a shriveling apple. Even in his first report on the high plateau country, Dutton stressed this inconsistency.

After several field seasons north of the Grand Canyon, Dutton's 1882 field assignment was to a very different region: the volcanic Cascade Mountains of Oregon and northern California. But something curious occurred: Powell suddenly delayed this assignment for a year while others mapped the area. At the time, nobody seems to have commented that Dutton had demonstrated extraordinary ability to organize mapping teams simultaneously with geological studies. The record becomes a trifle murky.

The Cascade Mountains assignment obviously was somewhat surprising. Dutton admittedly had little more than a textbook knowledge of huge volcanic mountains, but few in America had more. Some had visited Vesuvius, and Dana had studied volcanoes in Hawai'i and the south Pacific 40 years earlier during the Wilkes Expedition. Yet someone may have sug-

[3] Gilbert was perhaps a greater field geologist than either Dutton or Powell, but he abhorred administrative duties and was a less skilled writer. The cited publication was his 1877 report on the geology of the Henry Mountains (see list of Dutton's principal writings).

[4] Dutton's notable report on the geology of the high plateaus of Utah was published in 1880 (see the list of principal writings).

gested that study teams in volcanic areas should be headed by someone who had more than textbook knowledge of volcanoes, or a different agenda may have existed privately. In any event, Dutton spent the 1882 field season in Hawai'i instead of Oregon and California. And he was dispatched on very short notice.

Dutton after His Hawaiian Trip

During late 1882 and early 1883 Dutton completed his Hawaiian report and saw it published. He spent the 1883 and 1884 field seasons in northwestern New Mexico before proceeding to the Cascade Mountains in 1885 and 1886. From the former he produced the third volume of his great trilogy about the high plateaus of the western United States.[5] In 1884 he was elected to the National Academy of Sciences (then as now, a recognition of special professional merit) and he became director of the newly designated Division of Volcanic Geology of the U.S. Geological Survey. He was in great demand as a speaker, and early in 1884 he spoke out, expressing great confidence in the monarchy and legislature as well as the Hawaiian people.[6] Unfortunately he placed it in the context of their being "the finest and most intelligent race of dark-skinned people of the world"—today obviously a politically incorrect phrase. And he gave the missionaries of the 1820s more credit than do many revisionist historians. But this was the bad old 1880s, and he seems to have been far in advance of his times as a social philosopher.

The year 1885 was especially notable for Dutton's sounding the depth of Oregon's Crater Lake and his correct conclusion that it occupies a typical volcanic caldera. Subsequently he urged setting aside the region for what eventually became Crater Lake National Park. But he never produced a monographic report on the Cascade Mountains. He spent autumn 1886 and all of 1887 in charge of investigating the great Charleston (S.C.) earthquake of August 31, 1886. He ultimately produced a monographic report on that disaster, which ranks with his three plateau reports and the present volume about Hawai'i.[7]

Meanwhile, the political climate in Washington, D.C., had changed. Grover Cleveland became president in 1885. He opposed annexation of Hawai'i and must have welcomed "Hawaiian Volcanoes" and Dutton's

---

[5] This third volume of the high plateau trilogy was his 1885 *Mount Taylor and the Zuni Plateau* (see list of principal writings).

[6] Dutton's lecture on "The Hawaiian Islands and People" was delivered at the U.S. National Museum on 9 February 1884 under the auspices of the Smithsonian Institution and the Anthropological and Biological Societies of Washington. A 32-page pamphlet version of its text was published by Judd & Dettweiler.

[7] The principal report on this earthquake is Dutton's 1889 "The Charleston Earthquake of August 31, 1886." (See list of Dutton's principal writings.)

1884 speech. But changes in direction by the new Congress caused rifts in the U.S. Geological Survey. In October 1888 Congress appropriated $100,000 specifically to begin a United States Irrigation Survey. Dutton had experience in this field, and it was logical for him to be placed in charge, first of its hydraulic and engineering division, then the entire Irrigation Survey. Unfortunately it led to disaster when Powell diverted Irrigation Survey funds to topographic mapping.

Pro-annexation Benjamin Harrison became president in 1889. In that year, Dutton resurrected and popularized the largely forgotten concept of isostasy: the up-and-down movement of parts of the earth's surface in response to weighting down and to deloading. Among some of his fellow geologists, it is his best-remembered contribution. But it was not enough. In 1889 Dutton's annual "leave" from the Ordnance Corps was not renewed. He left the U.S. Geological Survey and returned to active duty. But it is difficult to agree with one biographer who termed this a mark of disgrace. Promoted to major, he became commandant of the San Antonio (Texas) Arsenal where official duties were light. In April 1891 he was dispatched to Nicaragua for what now would be termed military intelligence. At this time, Congress was considering the relative merits of Panama and Nicaragua for an interocean canal, and concerns had been raised about volcanoes and earthquakes in Nicaragua. (In 1902 the Nicaragua Canal Commission noted that "no one is better qualified to speak on this subject.") Writing as an engineering geologist he promptly reported to the Army's Division of Military Information that the project was staggering in scale but appeared feasible. In a long letter written in response to a request from the Nicaragua Canal Construction Company (to which Dutton and the U.S. Army were indebted for detailed maps and profiles he had acquired), he provided much the same findings and conclusions. Throughout the next decade, both found their way into a series of Congressional documents.[8] A century later, the Nicaragua Canal proposal remains viable. The Panama Canal is too narrow for today's supertankers and superliners.

Little else is on record from Dutton's decade in San Antonio. He became especially friendly with George W. Breckenridge, a well-to-do banker and industrialist. With Breckenridge he somehow traveled to Italy, Egypt, and India, but apparently did not continue around the world. No one has found any further mention of Hawai'i, where ships from Asia and Australia stopped en route to America. After 10 years in Texas, he returned to Washington, D.C., as assistant to the chief of the Ordnance Corps. But he was

[8] Dutton's 24-page military intelligence report was entitled "Report on the Nicaragua Canal." It was filed in 1891 but generally is dated March 18, 1892, the date the Secretary of War transmitted it to the U.S. Senate. His letter appeared in the 1891 prospectus of the Nicaragua Canal Construction Company and in various Congressional documents, most notably as Exhibit C in the 1902 Report of the Nicaragua Canal Commission.

increasingly enfeebled (probably by chronic tuberculosis, to which he must have been exposed repeatedly in the course of his fieldwork).[9] He retired in 1901, but for a time he was able to keep pace with exciting new landmarks in geology. In 1904 he published a much-praised book on earthquakes as effects of geological processes, then a startling new concept.[10] In 1906 his last major publication concerned the profound changes the brand-new discovery of radioactivity was bringing to the science of geology, with special reference to the origin of volcanic heat.[11]

In retrospect, the world should hail Clarence Dutton for much: for extraordinary explorations and technical geological contributions; for uncomplaining acceptance of remarkable physical exertion, hardships, and danger; for the enthusiasm and skill with which he brought his exceptional geological observations to life on paper; and for his perceptive exposition of both the volcanic wonders of Hawai'i and the innate strength of the Hawaiian people. This twenty-first century republication of one of his superb geological travelogs is but a small tribute to a remarkable man.

---

[9] Most of Dutton's biographers are silent about the cause of his death. Writing an obituary for the National Academy of Sciences, Chester Longwell attributed it to "arteriosclerosis"—then a euphemism for what is now known as Alzheimer's disease. The praise given his 1904 book on earthquakes indicates that this was not the disabling chronic problem that began several years earlier and progressed relentlessly, however. His son called this "influenza" but this is an acute short-lived disease unless complications develop. Bacterial empyema (a pus-forming infection between the lung and the chest wall) was one such complication but was relatively uncommon. Producing somewhat similar symptoms but much commoner was tuberculosis, to which Dutton must have been exposed in the primitive sanitation of the battlefield camps of the Civil War, in his field camps of the 1870s and 1880s, and probably during his visits to Egypt and India. At the turn of the century, tuberculosis usually was not diagnosed until the late stages of the disease, which generally occurred late in life in persons as robust as Dutton. Until mid-20th century, when effective treatment became available, it was broadly considered a social stigma, and such terms as "influenza" were used instead.

[10] Dutton's 1904 book on earthquakes received special praise in biographies by fellow geologist J. S. Diller.

[11] "Volcanoes and Radioactivity," 1908.

Appendix C
Principal Writings of Clarence Edward Dutton

1871 The causes of regional elevations and subsidences. *Proceedings of the American Philosophical Society* 12:70–72.

1874 A criticism of the contractional hypothesis. *American Journal of Science*, 3rd series, 8:113–123.

1876 Critical observations on theories of the Earth's physical evolution. *Geological Magazine*, 2nd Series, 3:322–328, 370–376. Also in: *Pennsylvania Monthly*, 7:354–378, 417–431. Abstract, *American Journal of Science*, 3rd series, 23:142–145.

1877 Report on the lithologic characters of the Henry Mountain intrusives. In *Report on the Geology of the Henry Mountains*, edited by G. K. Gilbert, 61–65. U.S. Geographic and Geological Survey of the Rocky Mountain Region. Washington, D.C., GPO. 2nd ed. 1880, The Intrusive Rocks of the Henry Mountains, pp. 147–151.

1878 Irrigable lands of the valley of the Sevier River. In *Report on the Lands of the Arid Region of the United States*, edited by J. W. Powell, 128–149. Washington, D.C., GPO.

1879 Geological history of the Colorado River and plateaus. *Nature* 19:247–272.

1880 *Report on the Geology of the High Plateaus of Utah*. U.S. Geographical and Geological Survey of the Rocky Mountain Region. Vol. 32, with atlas.

1880 The causes of glacial climate. *Bulletin of the Philosophical Society of Washington* 2:43–48.

1880 On the Permian Formation of North America. *Bulletin of the Philosophical Society of Washington* 3:67–68.

1881 On the cause of the arid climate of the western portion of the United States. *Proceedings of the American Association of the Advancement of Science* 30:125–128.

1881 The excavation of the Grand Canyon of the Colorado River (abstract). *Proceedings of the American Association for the Advancement of Science* 30:128–130. Also in: *Science* 2:453–454.

1882 The Physical Geology of the Grand Canyon District. In *2nd Annual Report of the U.S. Geological Survey*, 47–166. Washington, D.C., GPO.

1882 Review of *Physics of the Earth's Crust* by Osmond Fisher. *American Journal of Science*, 3rd Series, 23:283–290.

1882 *Tertiary History of the Grand Canyon District.* U.S. Geological Survey Monograph 2, with atlas. Reprinted 2001 by University of Arizona Press with a new foreword by Stephen J. Pyne.

1883 Recent explorations of the volcanic phenomena of the Hawaiian Islands. *American Journal of Science*, 3rd Series, 15:219–226.

1883 Petrographic notes on the volcanic rocks collected by W. H. Holmes in the Yellowstone National Park. In *U.S. Geographical and Geological Survey of the Territories, 12th Annual Report*, pt. 2, 57–62. Washington, D.C., GPO.

1884 Hawaiian volcanoes. In *4th Annual Report of the U.S. Geological Survey*, 75–219.

1884 The Hawaiian Islands and people. A lecture delivered at the U.S. National Museum under the auspices of the Smithsonian Institution and of the Anthropological and Biological Societies of Washington. February 9, 1884. Washington, D.C., Judd & Detweiler.

1884 The Geology of the Hawaiian Islands. *Bulletin of the Philosophical Society of Washington* 6:13–14.

1884 The effect of a warmer climate upon glaciers. *American Journal of Science*, 3rd Series, 27:1–18.

1884 The volcanic problem stated (abstract). *Bulletin of the Philosophical Society of Washington* 6:87–92.

1884 The basalt fields of New Mexico. *Nature* 32:88–89.

1885 The volcanoes and lava fields of New Mexico (abstract). *Bulletin of the Philosophical Society of Washington* 7:76–79.

1885 The latest volcanic eruption in the United States. [Lassen Peak, 1883]. *Science* 6:46–47.

1885 Mount Taylor and the Zuni Plateau. In *6th Annual Report of the U.S. Geological Survey*, 106–198. Washington, D.C., GPO.

1886 Crater Lake, Oregon, a proposed national reservation. *Science* 7:179–182.

1887 The submerged trees of the Columbia River. *Science* 9:82–84.

1887 The Charleston earthquake. *Science* 10:10–11, 35–36.

1887 [with E. Hayden] Abstract of the Results of the Investigation of the Charleston Earthquake. *Science* 9:489–501.

1888 On the geologic nomenclature in general and the classification nomenclature and distinctive characteristics of the Pre-Cambrian Formation and the origin of serpentine. In *4th International Geological Congress American Committee Reports*, 71–73.

1888 On the depth of earthquake foci. *Bulletin of the Philosophical Society of Washington* 10:17–19.

1888 [with S. Newcomb] The speed of propagation of the Charleston earthquake. *American Journal of Science*, 3rd Series, 35:1–15.

1889 On some of the greater problems of physical geology. *Bulletin of the Philosophical Society of Washington* 11:51–64.

1889 The Charleston earthquake of August 31, 1886. In *9th Annual Report of the U.S. Geological Survey*, 203–528. Washington, D.C., GPO.

1891 The crystalline rocks of northern California and southern Oregon. In *Compte Rendu, 4th Session, International Congress of Geology*, 176–179.

1891 *Report on the Nicaragua Canal. Report to the U.S. Army Division of Military Information.* 24 pp. Transmitted by the Secretary of War to the U.S. Senate March 18, 1892. Senate Misc. Doc. 97, 52nd Cong., 1st sess. Serial 2904. Also in House Rept. 2126. 54th Cong., 1st sess.

1891 Letter to the President of the Nicaragua Canal Construction Company, apparently first published in 1891 prospectus of that company as "Volcanoes and Earthquakes, Nicaragua and Costa Rica," pp. 73–78. Also pp. 55–62 in J. G. Walker, Lewis A. Haupt, and Peter C. Hains. 1902. *The Interoceanic Canal: Report of the Nicaragua Canal Commission*. Senate Doc. 357, 57th Cong., 1st sess. Serial 4245.

1904 *Earthquakes in the Light of the New Seismology*. New York, G. P. Putnam.

1906 Volcanoes and radioactivity. *Journal of Geology* 14:259–268. Also in: *Popular Science Monthly* 68:543–550. First published as a 12-page pamphlet printed by Englewood (N.J.) Times.

Dutton's U.S. Geological Survey administrative reports provide additional autobiographical information. These appear in the 1st, 2nd, 3rd, 4th, 6th, 7th, 8th, 9th, and 10th annual reports of the Survey, 1880–1890. Only in the *10th Annual Report* does Dutton's administrative report carry a specific title: "United States Irrigation Survey 1888–1889."